My Life with the C-130 Hercules

ROSS HOLDEMAN

🔥 Light Messages

Copyright © 2015, by Light Messages Publishing
My Life with the C-130 Hercules
Ross Holdeman

Published 2015, Light Messages Publishing
www.lightmessages.com
Durham, NC 27713 USA
SAN: 920-9298

ISBN Paperback: 978-1-61153-180-0
ISBN Hardcover: 978-1-61153-762-8
ISBN Ebook: 978-1-61153-747-5

ALL RIGHTS RESERVED

No part of this publication may be reproduced, stored in a retrieval system, or transmitted in any form or by any means, electronic, mechanical, photocopying, recording, scanning, or otherwise, except as permitted under Section 107 or 108 of the 1976 International Copyright Act, without the prior written permission except in brief quotations embodied in critical articles and reviews.

This manuscript was written by Ross Holdeman in his personal capacity. The opinions expressed in this book are the author's own and do not reflect the views of the publisher. This book has been published at the request of the author as a personal memoir for his family and friends and not as a public historical record. The publisher has made reasonable attempt to verify the text for historical and factual accuracy but the nature of the accounts are such that much can only be known by the author.

PREFACE

Many great stories have probably been told, and I could make my contribution as well, but I hope that none present lose sight of their true significance. Such a wealth of remembrances can only occur when one has done much, contributed to many, and has as true result, earned the respect and admiration of those present.

You have, over the many years of our friendship and working together, enriched my life beyond my ability to express. You have been a patient teacher, loyal advisor, guardian of the mission, caretaker of the flock, and a great person to share time with. We all owe you a great deal, which we probably will never be able to repay.

You leave a legacy of contributions to aviation, safety, and especially the many who have operated the C-130 throughout the world. Through you, a large portion of the much enjoyed Lockheed reputation has been planted, nourished, and harvested. You have been a good gardener!

Jerry G. Valentine
Lieutenant Colonel, USAF
Deputy Directory, Quality Assurance

ONE

On June 19, 1951, I started working for the Lockheed-Georgia Company (GELAC) division of Lockheed as an aeronautical engineer. In October of 1952 I was transferred to the C-130 Project Engineering Department (Dept. 72-05) after Lockheed was contracted to develop a tactical airlift aircraft for the United States Armed Services.

As the Prototype Lead Project Engineer Willis Hawkins had guided the design and prototyping of the airplane at the Lockheed-California Company (CALAC) plant but production engineering and manufacturing were to be done in Georgia. By June of 1952, full production engineering responsibility was transferred from CALAC to GELAC.

I worked in the Landing Gear Fuselage and Power Plant Design groups and, as my first love, was working hands on with the "hardware." I was assigned to the production floor as a liaison engineer assigned to the Production Liaison Group when 3001 (the first production C-130) was moved from the production floor to the flight line. On February 24, 1955, I was assigned to follow it as a liaison engineer at the flight line. The first flight of the 3001 was made on April 7,1955. The flight was uneventful;

it reached an altitude of 20,000 feet. All controls, as well as the landing gear and flaps, were retracted and extended, the aircraft was slowed to stall buffet, and the flight lasted one hour and 34 minutes. All in all, it was a very successful, uneventful flight with all the preplanned maneuvers accomplished.

During landing of the third flight and reversal of the propellers, a quick disconnect fitting came apart and fuel ignited. The end result was that the left wing was destroyed before the fire could be extinguished. Although there was a lot of finger pointing, it was finally and firmly established that it was against Lockheed policy to use quick disconnects on fuel lines in the area of engines. That had been done, in this instance, only as a result of an Air Force directive. Needless to say, no more quick disconnects were installed on engine fuel lines.

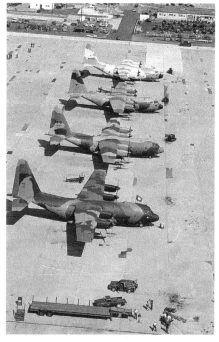

In July of 1956, the 6th production of the C-130, S/N 3006, went to Pope Air Force Base with a mixed crew, part Air Force and part Lockheed, to do paratrooper drop tests. On the 18th of August the Main Landing Gear (MLG) only partially extended when coming in for a landing. The runway had foam applied and the plane came in wheels up. We on the flight line were not surprised as we had always had trouble with the MLG and were always told "if it is adjusted correctly, it works," which was a true statement. The problem was that after an undetermined number of cycles the wear was such that the MLG was no longer adjusted correctly.

With an "I told you so," I and the Group Engineer for the Landing Gear Group were sent to Pope Air Force Base to investigate and get the aircraft ready for a ferry flight to GELAC.

I was not very diplomatic in pointing out the reason for the malfunction with numerous "I tried to tell you" comments. It took some time for me to overcome my unnecessary finger pointing, but the Group Engineer and I eventually became very good friends.

Much to my satisfaction changes were made to the Main Landing Gear, and to my knowledge, there were only three other instances of gear-up landings. I investigated all three and found gross maintenance neglect. However, this is where I learned that in a large company you never ever say "I told you so." I won't go into the details, but it was some time before I was allowed off the flight line to even visit the C-130 Project Engineering area. All contact had to be either written or by phone conversations.

I learned that large engineering organizations are setup so that the responsibility can never be isolated to an individual. This was something that was completely foreign to me at the time, but as time passed I learned to live with it and then to understand it. You might be criticized or even fired, but you could never be held responsible for your decisions.

Due to problems from day one with the Curtis Wright propellers, the United States Air Force (USAF) issued a contract to the Aeroproducts Division of Allison to build propellers for the C-130. In July of 1956, Aeroproducts propellers were installed on 3,006 C-130 aircraft. Any C-130s with Curtis Electric propellers were grounded in November of 1956 until Aeroproduct propellers could be installed in them. However, production of the C-130 slowed but did not cease. They were built, flown one time, then after the Air Force purchase form, DD250, was signed, the propellers were removed, and the aircraft was stored.

At one time we had 49 aircraft in a 50 aircraft oblong circle on an abandoned runway at Dobbins Air Force Base, called the "runway egg." Once a week the planes were each pulled forward one aircraft length to rotate the tires. Eventually, these aircraft were rotated back through the production line to have Aeroproduct propellers installed as well as all the structural and system changes dictated by flight and static tests. These structural changes were incorporated system-wide so that all aircraft were configured the same. This meant that these aircraft, as well as new aircraft, entered flight stage at an unanticipated rate. During one month we delivered 22 new and modified aircraft to the Unites States Air Force (USAF).

As time went on, the harried pace did slow to a more manageable rate. During this time I discovered that the head of the USAF Flight Acceptance group, a Colonel Little, had been an underclass-mate of mine during my cadet days. This gave me a bit of a leg up, and I started going on flights to try to locate the causes of in-flight problems that could not be duplicated on the ground. This allowed me to eventually get a lot of C-130 "stick time," much to my enjoyment. I'll never really know how much this willingness to fly and trouble shoot helped my career.

It almost came to a halt one day when there was an in-flight problem with the aft cargo ramp door locking in the open up position with the aircraft flying at 200 knots, which was a contractual requirement. I went on a flight, wearing a parachute in case things went wrong, to attempt to see what was going on. With the aft cargo door opened, I was standing on the ramp trying to see what was going on. When we hit a thermal I lost my balance and fell overboard. I recovered from the shock and realized that I was  airborne without being in the airplane. I looked around and pulled the rip cord on my parachute. The airplane circled me, radioed the plant, and a chopper was dispatched.

After I landed on the ground, I simply gathered up my chute, looked around, and saw a farmer who had been plowing come running over. In a few minutes I heard a chopper and, sure enough, he landed. I climbed aboard and off we went back to Lockheed. There was quite a fuss made, including a physical examination. Lockheed and I were mum on the event, but the next issue of the Marietta newspaper revealed many facts that I never knew, apparently from an interview with the farmer. I was always glad that Lockheed Public Relations dummied up with "no comment."

From that time forth, anytime you were in the cargo compartment and the ramp/aft cargo doors were opened you wore a safety harness tied to the interior structure of the fuselage that precluded falling out of the airplane. This safety practice was adopted by the USAF and as far as I know is still in effect, which I think is a good idea.

1957 was a very harried year. By this time it had become evident that the C-130 filled a gap in the Air Force air lift ability. During this year we built working demonstration versions that could land with skis on snow and ice or hard surface runways on wheels, and an in-flight refueling version for the United States Marine Corps (USMC). Also we started working on an upgraded version with Hamilton Standard Propellers that had larger engines which became the C-130B and developed a C-130A with different electronics and a few other subtle changes for the Royal Australian Air Force (RAAF). We eventually delivered twelve C-130A aircraft to the RAAF.

All of this, in addition to a new production rate of three aircraft a week and a minimum delivery rate of four a week of the modified C-130s from the "runway egg," made a weekend off something that Flight Line personnel never even thought about. We were working twelve to fourteen hour days, seven days a week, and in those days there was no overtime for salaried employees. When we were about to drop, the flight line would shut down for a Sunday.

In the midst of all this, C-130B production was starting up and the propellers from a new manufacturer that had been specified for the C-130B were causing a lot of grief. I was called in to a meeting between Engineering, Production, and upper management. After a lot of discussion, I was asked if I could get the propellers straightened out. I replied that "I probably could but I couldn't and be nice to the supplier". With that the Vice President slammed his hand down on his desk and said, "Dammit! I didn't ask you to win a popularity contest, I asked if you could get these problems straightened out."

With that I started commuting between the Atlanta and Hartford, Connecticut airports. I could leave Atlanta around 4:00 A.M. and be at the propeller manufacturer's plant just after work started for the day. Some sessions there were rougher than others, but progress was being made. One evening while standing in line at the Hartford airport to board a flight to New York and then change planes for a flight to Atlanta, I heard a voice about two people back say, "Well, I guess the plant is happy now that the Spherical SOB Holdeman has left."

There was a bit more banter between the two guys behind me. I thought about it a bit and then curiosity got the best of me. I stepped out of line and stepped back to the two people and said, "Pardon me, but I overheard some of your conversation and something has me curious. I thought that I had heard of every kind of a SOB there was, but I never heard of a Spherical one — what is that?"

This one guy made a ball with his hands and said, "Well, a sphere is perfectly round so it is the same from any angle you look at it".

I sort of smiled and said, "Thanks. By the way — my name is Ross Holdeman." Instead of being upset, I was rather pleased that I was getting their attention.

I never told anyone at the plant about the conversation I overheard and never mentioned it until about six months later. We were in a rather heated meeting at Lockheed with the propeller manufacturer and I thought things needed relaxing a bit. So I told my story. Instead of calming the situation, my story had the opposite effect. The propeller manufacturer got all pushed out of shape, demanding to know who made the comments. I refused to tell them. They even went so far as to get my vice president to call me in and ask. I told him that I had taken it as a compliment that I was doing the job he had expected me to do. I told the Vice President that the man who made the comment was an excellent engineer on the propeller and added, "You can bet that every time his desk phone rang for some time afterwards he probably wet himself just a little wondering if I had reported him."

I'm sure his employer could rest assured that he would never be that indiscreet again. Over the years this guy and I actually became good friends but that incident was never mentioned and I have never revealed his name.

My efforts must have been effective as that propeller system was used on new production Hercs for 40 years on more than 2,100 airplanes. Naturally, over the years improvements and changes took place, but the basic propeller system remained the same.

On September 2, 1958 we lost our first C-130. It was a specially-modified aircraft, part of the "Sneaky Peat" squadron that was to fly very close to the USSR border in order to listen in on Soviet military communications using special antennas, radios, and recording equipment. On the above date, the Soviets managed to "bend" a radio navigational aid and "suckered" the aircraft over the border into Soviet air space where fighters were waiting and shot it down. In later years I became friends with a pilot and air traffic controller from that squadron.

In the early 60s I was sent to Berlin, Germany. The Canadians had an oxygen fire on a C-130 there and I was to survey the damage. After my survey, I boarded a Pan Am flight. While taxing out for take-off, the plane stopped and the passenger door opened. A USAF Captain appeared and using the on board intercom asked for Mr. Holdeman to please identify himself. This didn't set very well with me so I remained quiet. After the third request, the Captain informed all the passengers and crew that "this plane is not departing until Mr. Holdeman identifies himself." I gave up and did so. He informed me that the Colonel wanted to see me and my baggage was being removed from the plane.

When we got to the Colonel he informed me that, because I had been designated as the C-130 technical contact for the USAF by Lockheed, the USAF was evoking their right for assistance, and Lockheed had been so advised. It seemed that with this being the height of the Cold War, the United States had kicked out a highly-ranked Russian official from the Russian Embassy in Washington as a spy and Russia had promptly responded, giving 48 hours for the departure of a U.S. special envoy and his family from Moscow. The USAF had set up a C-130A flight to remove the official, his family, and household belongings. However, the catch was that the plane would be furnished with a Russian navigator and allowed a USAF crew of only four. As I had some experience in the right seat of the C-130A and a reputation for "moving a plane — legally or otherwise," I was to be co-pilot with the responsibility of getting the plane to Moscow and back, whatever the conditions. To say I was apprehensive would be an understatement, but I did not have a choice.

The flight went extremely well and we delivered the high-ranking official and his family back to Turkey with no problems. We did have a bit of fun on the trip from Moscow to Incirlik, Turkey, shaking up our Russian navigator by shutting down an engine just for fun.

From Turkey I flew commercially back to Atlanta where the CIA immediately debriefed me over a couple of days about what I had learned. Lockheed security and an upper management person were also present. It all ended with several warnings about not revealing anything about the incident.

It wasn't long after this incident that I was called into the Chief Engineer's office and told that he was going to step out into the hall, but someone would come in to talk with me. He went on to say that no one would order me to do anything, but it was Lockheed's policy to cooperate fully with these people.

The CIA had lost two C-130 aircraft on "spook" missions. I knew of one such incident and had sounded off around Lockheed that had the CIA known more about how to operate the aircraft, they could have brought it home rather than destroying it. Thus began my association with the US "Spook" organizations.

I never knew who paid for what; I only drew my normal Lockheed salary and submitted the normal expense statements. I would make a "normal" trip for Lockheed and when it was an overseas trip, upon my return, I would usually be contacted by unknowns and queried a bit more extensively. At times I would be called in and handed travel orders by Lockheed.

It was my primary job to be sure that the airplane returned, regardless. I brought one back with only three main landing gears. I brought a couple back full of holes (not counting Vietnam) and one back with no hydraulic pressure — that was no fun. I almost had to clobber the crew to keep them from jumping.

I failed one time and that was in the Iranian desert during the ill-conceived effort to rescue the U.S. Embassy personnel being held hostage by the militant Iranians. This operation was so ill-conceived and rushed in an effort to re-elect Jimmy Carter to the Presidency that it was doomed to failure from the start. The loss of the C-130 was the result of the "Aerial convoy" landing in Iran on a salt bed flat to refuel the choppers. One chopper took off and hit a C-130 with its engines idling resulting in a big ball of fire. That trip resulted in my first of two rides on the supersonic Concord. I never really knew why but upon my return to Frankfurt, the local Lockheed representative handed me a ticket to Paris, then to Washington on the Concord, and finally home on Delta. Naturally, there was a long session in the round "Secure Room" the next day.

Over the years I was involved in twelve to fifteen such operations. Most were routine trips, like being sent to Beijing to assure there were no hang-ups for the Hercs that carried the TV satellite to broadcast President Nixon's visit to China in February of 1972. It was an interesting trip but my services were never required as all went as smooth as silk.

There were a couple of flights into areas where we had not been invited that resulted in the airplane receiving a few holes from unfriendly fire. One such incident led to a bad landing in a remote, dry riverbed that tore the front main landing gear off on the right side. I was able to make a "lash up" and we came back to friendly territory with only three main landing gears.

In 1963 Lockheed was trying to determine what had gone on during a specific C-130 Navy mission, how the aircraft had been modified and how it worked, without satisfaction. Lockheed applied to the Navy for a Lockheed person to go on one of the missions. As luck would have it I was the one approved, but I was required to go on a specific mission. When Lockheed received the approval, I was in Canada with Pacific Western. I was just leaving and was at the Edmonton International Airport when I was paged. Pacific Western had discovered the first wing crack in a commercial aircraft. I returned to Edmonton Industrial Airport, inspected the wing, and called Lockheed requesting a structural fix scheme. Within an hour Lockheed called back and told me they were sending a structural engineer to fix the commercial aircraft and I was to get to the Patuxent River Navy Base in time to make the pre-trip briefing.

Six days later I was back at Lockheed in the round "Secure Room" for my debriefing on the Navy mission being given all sorts of warnings — all aircraft performance flight maneuvers were on "a need to know basis" and there would be "dire consequences" if any further discussion occurred between Lockheed personnel. As a result Lockheed was able to make several recommendations as to modifications that would make the plane a better platform for the mission. Much later we discovered that most of our recommendations had been incorporated.

One of my biggest problems was going along with the official cover story to my wife, Louise, when I was going on one of those missions. The Navy even had a Lockheed Vice President write her a letter thanking her for my service to Lockheed. To this day it bothers me that I must guard my conversation if it gets into the wrong area. I would dearly love to tell her everything, but should I do this and she says the wrong thing to the wrong person, it could put her at risk. There are things I know about a couple of missions I was on that some people would still go all out to protect against becoming common knowledge. I'm sure glad I don't do much talking in my sleep — only twice has Louise had to wake me when I was talking.

However, not too long after I retired and was doing some consulting, I received a call to come to San Antonio for a job discussion. There I was met by a couple of people at the hotel and after a bit of discussion we went to the bar for a drink. After that things get all mixed up in my memory. They somehow convinced me that I was in Macon, Georgia and kept trying to get me to jump out of a hotel room window. I finally called Louise, how I don't know, and told her I was in the Macon Hilton on the 10th floor and to please come and get me because someone was trying to kill me. She got our pastor to drive her to Macon at 2:30 in the morning only to discover that there was no 10th floor in the Macon Hilton. I don't know what I had been given or even how I finally got away.

The next thing that I vaguely remember was being all confused the next morning, but finally getting it together enough to get to the airport, board a plane, and fly home. I was still so confused that I actually got lost driving from the Atlanta airport to my home. It was about a week before I didn't have bad blank periods.

A couple of weeks later, I received a call that it would be appreciated if I would come down to Lockheed for a meeting in the round "Secure Room." The night before I spent the night composing three copies of a letter that were mailed to a federal judge, an attorney, and a minister all with instructions to open the letter and publicize the contents if I died or disappeared suddenly or, if convinced that I had died naturally, to please destroy without opening. On the way to Lockheed I mailed one in Waleska, one in Canton, and one in Marietta. I evidently was being followed as at the meeting when I told them what I had done, one of them confirmed that I had done just that (and this is the clean version of that meeting). The meeting ended with a plea to let bygones be bygones and to continue not divulging certain information.

I know that it has been a long time but frequently I have a visit "just to remind me." As late as 2002 we had one of these visits. The man was also a personal friend and as this was his second trip, I finally told him that he was welcome back anytime as a friend but to never darken my door in an official capacity again. So far it has held. How long I don't know — I am sure that some people will be relieved when I kick the bucket.

As my knowledge of the C-130 continued to expand, and to some extent my reputation in the C-130 community, and the C-130 was being delivered to more and more foreign customers, my travels became more and more frequent both for legitimate customer calls and otherwise. Then my duties were expanded when I was made Chief Engineering Accident Investigator. As Lockheed gained the right to sell to foreign governments (regardless of technical ability) and with the Vietnam war in full swing, the accident rate went up. However, it was not until October of 1965 that Lockheed became actively involved in accident investigations.

By 1968, Lockheed's expertise in accident investigation of the C-130 had proved to be of such technical assistance that the United States Air Force Military Airlift Command Headquarters had changed the third item on the "to do" checklist to: When a C-130 crashes, call Lockheed and request Ross Holdeman's assistance. Work in this area actually enhanced my reputation with the C-130 operating units as well as with Lockheed.

Sometime in the early 1960s I was at a USAF C-130 base on a problem, (I think that it was Pope Air Force Base). That evening at the Officers Club bar a general bull session started between the C-130 crews and myself. I kept hearing, "that airplane will fly you out of anything you can get it into" or words to that effect. After about the third time I heard this, I became alarmed and started telling them that the airplane would kill them. Upon returning to Lockheed, I requested a meeting with the C-130 project's Chief Engineer, Chief Safety Engineer, and Vice President. Several others attended the meeting as well. I did a hard sell that we had built such a well-performing airplane that it was creating overconfidence among the flight crews, and if this overconfidence was not replaced with proper respect for the C-130, we were going to kill a lot of people. The reactions were very mixed from pride, to disbelief, to actual shock. I was asked, "What would you suggest?"

I replied, "Somehow we must get the word out there of just what I kept saying in the O-Club —that airplane will kill you." I didn't have any answer for how this should be done.

Actually, we were in a most difficult position. On one hand there was the pride in the airplane and the crews belief in it, and on the other hand everyone in the meeting knew I was right.

After several brainstorming meetings, kicking around ideas, and gnashing of teeth it was decided that Lockheed would form a C-130 Safety Briefing Team and try to get funding from the USAF. Lockheed enlisted support from Allison and Hamilton Standard and the team was born. They hit the road immediately.

My last trip for Lockheed was to take the team to Point Magu Naval Air Station (NAS) for a safety briefing. The C-130 Safety Briefing Team actually survived until 1995 still consisting of a Lockheed C-130 project representative, safety engineer, C-130 pilot, and a representative from both Hamilton Standard and Allison. They traveled all over the world visiting all U.S. military, civilian, and foreign operators of the C-130. This eventually evolved to where it occupied about a quarter of my time. As the United States Air Force and United States Marine Corp Air Guard and Reserve began flying C-130s, a great number of weekends were used briefing these units as well. (Remember — there was no overtime for Lockheed's salaried employees).

During the engineering build-up for the C5A, many newly employed and contract engineers were arriving for work "at their convenience." The word went out from the Chief Engineer that anyone with Department 72XX (the Engineering Department) on their badge arriving after 7:30 A.M. (our normal start time) would have to punch a time card at the guard house and give it to the guard, who in turn would deliver it to the Chief Engineer.

I don't know how effective this procedure was, but one evening after work Delta Airlines called and said that they had a Herc hung up in New Orleans and they had a pass waiting for me at the gate for their flight to New Orleans leaving in about one hour and to please be on it. Off I went. I lucked out and actually lashed up the problem and we brought the flight on to Atlanta where a permanent correction could be made. I went home, showered, and went to the plant arriving about 7:40 A.M. The guard saw 72-05 on my badge and handed me a time card to punch in on the time clock. I looked at it, turned around, walked out, got in my car, and drove over to the Holiday Inn, ordered a cup of coffee and called the Project Engineer. I told him where I had been all the night before, where I was now, and that I had never punched a time card in my life and was too old to start now. I offered to buy him a cup of coffee if he would come over and join me. He cracked up and said to come on back and he would meet me in the parking lot in a company car and bring me in that way. I never did punch a time card.

Sometime in the fall of 1966 we were developing the HC-130H with the capability of "snatching" a downed pilot out of the jungle, bringing him on board, and also refueling helicopters in-flight with the hose and probe system. This system utilized a wing-mounted hose and reel system that had been developed for the USMC KC-130 in-flight refueling of fighter aircraft. With the "dirty configuration" of the HC-130H and the requirement of the hose reel drogue to be fully open at 125 knots indicated air speed (KIAS), we were having our problems.

Two

I was going to California, doing double duty. I was working with Sergeant Fletcher, the suppliers of the hose and reel assembly to correct a reel response problem and was to attend a Navy safety conference at Lockheed California Company (CALAC). My chief engineer was coming out to attend the conference and was staying out to visit friends over the weekend after the conference closed on Friday. I was going to take some modified hose reel parts with me on the red eye back home to install and test over the weekend. I picked up Chief Hercules Engineer Carol Dallas at his hotel, since I had a rental car. We attended the safety conference and CALAC had a dinner set up Friday evening after the conference. We were to attend and afterwards I was to drop Carol off at his hotel and catch the red eye back to Georgia.

During the pre-dinner cocktail party I met the Commanding Officer of the Marine C-130 unit stationed in Fatima, Okinawa. I knew that I was scheduled to take the Safety Briefing Team to Fatima the latter part of November so attempted to have a conversation. He was a bit high and kept sounding off at me that, "When you get there I think I will just send your a-s down South (Vietnam) for a visit with that detachment." About the third time he said this I stood up tall and said, "I guess you think that I wouldn't go," knowing full well what Lockheed's policy was about employees going into a combat area.

After dinner while I was driving Carol back to his hotel before heading to the airport, I told him about my encounter with the Marine Colonel, fully expecting him to forbid me going into Vietnam. He didn't respond and I glanced over at him. He was leaning against the passenger door. I asked if he had heard me. He responded, "Yes, and I was just thinking that if you went we could find out what the airplanes were actually going through." I was shocked.

Before I took the team to Okinawa, I tried every way I could think of to get a Lockheed order forbidding my going into a combat area. Unfortunately, Carol had beaten me to it, and I never could get anyone to say "no." Anyway, the latter part of November I took the team to Okinawa and upon arriving, the Colonel met the team, shook hands all around, invited us to the Officers Club for dinner, and handed me orders for detached duty to Da Nang (the base for the Marine C-130 detachment in Vietnam).

The briefing team did their briefings in Naha and Fatima and while the rest of the team headed back to the States, I climbed aboard a Marine C-130 headed to Da Nang. On the final approach I had a rude awakening; the crew pointed out some tracers slightly off to our right. When we got on the ground, I was taken to supply for outfitting with a cot, blankets, and a side arm. I told the sergeant that I was a civilian and didn't need the side arm. His response was, "The hell you don't if you want to live".

I spent seven days there sleeping about four hours a night. Otherwise I was going with the airplanes, learning what I could. I was assigned two warrant officers as escort; they took turns. The deal was that I could photograph anything unless I was told "no".

When I returned to Okinawa I had twelve, 36 exposure rolls of 35mm slides and 600 feet of Super 8 movie film. After returning home I had them processed and finally culled the 35mm slides down to about 200 and edited the movie film down to 225 feet. When I debriefed to the Project Engineer and told of being on board a flight that got hit and started a fire and a few other instances he was really shaken up about allowing me to get in harm's way and told me, "Talk to no one about this until I can get some guidance and I will get back to you." The next day he asked me to bring in the slides and movie film. When I pointed out that the movies were Super 8 and Lockheed didn't have that size projector, he told me to bring in mine. The slides and movie were first run before the Executive Committee who picked an audience to watch them one afternoon in one of the theaters. It was pretty obvious that some were people were upset by what they saw.

I did have some fun taking the pictures. Just before I turned in my combat flight gear and side arm, I had someone snap my picture standing beside the crew door wearing them. I "accidentally" included this slide along with airplane action shots. When I was doing the show for the GELAC Executive Committee I moved past the slide quickly. One rather slow Vice President asked me to go back, which I did, and he asked, "What is that on your side?"

I replied (I swear the devil made me do it), "A first aid kit."

He got up and walked toward the screen and said, "That sure looks like a gun to me."

I was very glad the rest of the committee broke up laughing as I couldn't hold it any longer.

Finally the Company President said, "Well, if you had been where he was, you would call it a first aid kit, too. Let's move on."

Eventually these pictures were taken to the corporate headquarters in Burbank and shown to the Lockheed Executive Committee and Board of Directors. What is so funny is that over the next eight years I made five additional trips to Vietnam but never went through more than the "normal" debriefing when I returned.

A few months later, I got the word that the Vice President for the C-130 program wanted to see me and I had an appointment at such and such a time. I hauled buggy into the Project Engineer's office to tell him, and he told me we had a meeting with the Vice President all set up. Off we went at the appointed time. It seemed that Lockheed and the United States Marine Corps (USMC) were having a big argument. Wing cracks were showing up and the USMC wanted Lockheed to fix them as a design defect. This disagreement had progressed to the point of going before the Department of Defense (DOD) Arbitration Committee. The Vice President was going to represent Lockheed before this committee and wanted to take about two minutes out of my movies in support of Lockheed's position that the airplanes were being utilized beyond their specification limits and, as a result, was being subjected to unanticipated use and abuse.

The Lockheed Photographic Department cut the desired sections, spliced them together, and then realized that they did not have a Super 8 movie projector. I was sent home to get mine for the Vice President to use the next day. I was later told that when the Vice President ran about 180 seconds of that film and turned off the projector to continue his pitch, the President of the DOD Arbitration Committee interrupted him, turned to the USMC and said, "When are you starting contract price and schedule negotiations with Lockheed for repairs?" When everything was returned to me after the presentation, I found a spare projector light bulb in the box.

From time to time I would be called to a meeting of some section of the C-5A engineering project and asked some specific question relating to aircraft usage under combat conditions. All in all it was a feather in my cap. I did apply for and received a 50% pay bonus for hazardous duty for the seven days I spent in Vietnam. I will never know how much money Lockheed made as a result of having those movie clips and certain slides. I do know that they were used several times to settle arguments relative to warranties with the USAF.

Shortly after my trip to Vietnam I was called in to a meeting with the Project Engineer and the Hercules Vice President and was told that they wanted me to move into an office in the Project Engineering area and concentrate on commercial operators as much as I could, reporting directly to the Assistant Chief Project Engineer. I asked what other duties could I drop and was told routine flight line assistance. I didn't feel like this new change would make much difference since, with field trouble shooting, problems in operational units and trips with the Safety Briefing Team, I had been doing less and less routine engineering assistance work on the flight line anyway. In fact, in 1967 I had traveled 110 days. I did a "box move" to the assigned office in Project Engineering and spent the rest of my career assigned to the C-130 Project Engineering offices.

After a few months, additional engineers were assigned to commercial engineering and the decision was made to make it a formal engineering group. I had always assumed that if this ever occurred I would be the group engineer. When I was informed that we were a formal engineering group and one of the other engineers was the group engineer, I was very upset. I went steaming into the Project Engineer's office and before I could open my mouth he said, "I didn't do it."

I stormed out and into the C-130 Project Engineering Manager's office, right by his secretary and blurted out, "Why?"

Jack got up, went to his office door, closed it, and I sat down.

The first thing that he said to me was, "I didn't invite you to sit."

I jumped up and immediately realized that I had probably made a tactical error.

In a very calm voice Jack said, "I have three things to say to you: One, even I would not work for you because you are too d—n much of a perfectionist. Two, you are too good at what you are doing to move. I will see to it that you will never suffer financially, even if we have to come with new job titles for you, but you will never do anything except what you are doing now. Three, if you ever come into this office again without being invited, I will personally fire you and walk you out of the plant. Now get out of here."

We did remain personal friends and I finished my career doing the same job but went through numerous titles to qualify for pay increases. I discovered that I was actually making more than the group engineer I was assigned to on paper. I finally retired with the high-sounding title of Senior C-130 Engineering Specialist. As far as I know, I am the only one to hold that title at Lockheed, for what it's worth.

During my final 25 years with Lockheed, I traveled over 2,856 days. I journeyed within eight miles of the North Pole, to the South Pole, and around the world several times, visiting 109 different countries, some numerous times. At some gateway cities I became known by name. One time when I was returning from a C-130 crash investigation in Germany, I walked up to the Delta desk in Frankfurt and the Delta station manager greeted me with, "Well, Ross, I was just wondering when you would be in. I heard about the crash and assumed we would see you."

Lisbon, London, and Paris were the other gateway cities where I was known by name. In many instances it helped. One December 22nd I arrived in Lisbon form Cameroon, Africa with no reservation for a flight home. Fortunately, the Lufthansa station manager saw me and very frankly told me that all flights from Europe to the USA were at least 10% oversold. Then he said, "Let me see what I can do." I got home on the 23rd — all I really know is that I got a first class seat assigned and made it.

Naturally, over time, I was given assess to Delta, TWA, Pan Am, Lufthansa, Swiss Air, and American airport VIP Clubs, which were a step above the first class lounges. When I retired, I had just over a million miles credit in the Delta Frequent Flyer Club. Not all these were flown on Delta, but by arrangements most of the airlines would accept miles from other airlines if you knew the ropes on how to get it accomplished, and I found out how and used it a lot. Remember that this was in the days when frequent flyer miles were credited only as a result of miles flown — no miles were given for grocery, gas, or other purchases. After retirement, Louise and I made several extended overseas trips on these frequent flyer miles and I still have some in the bank.

In November/December of 1970 I was in Okinawa for a few days — I forget why — but I was to return to the Marine Corps Air Station (MAS) El Toro near Irvine, California on a Marine Hercules. Since it was that time of the year, the crew had done a lot of Christmas buying in Japan and was bringing it all back. We were to clear customs in El Toro where the custom agents were very friendly with returning Marines. However, in flight we got diverted to Hawaii and panic ensued. There was not enough money on board to pay duty on all the high value electronic Christmas presents that were aboard and everyone knew "Honolulu is tough." Then I got an idea — put everything that we can on the aft cargo door, and when we land in Honolulu and taxi to the ramp, open the ramp and aft cargo door and invite the agents aboard. The agents shook the plane fairly well and were glad for the cool air inside the plane with the ramp, aft cargo door, and all overhead hatches open for air circulation (or so they thought). I became an instant genius to that crew. I was not involved but know of one Colonel that stored several thousand dollars worth of Turkish rugs to furnish a rather large house under the cargo floor and brought them back.

    I know of another attempt to avoid Customs duty that didn't turn out so well. A crew returning from Scotland loaded the wing dry bays with cases of good Scotch whiskey and flew home to McGuire Air Force Base in California. When they landed and Customs came out, they observed liquid dripping from the wings. At first they thought that it was fuel leaking, but a little closer observation resulted in the officials doubling over with laughter — not one bottle had survived the outside air temperature of - 40 degrees. The officials decided that the loss of all that premium Scotch was a sufficient price to pay. Lockheed was contacted as to whether any aircraft damage had occurred. I went out and supervised the flushing and re-treating of the dry bays after all the debris had been removed to prevent any corrosive action. There sure were a lot of "sad sacks" around that base.

    In June of 1968 I was in Iran on some problem and had a new experience while there. The Shah of Iran was giving a beautiful horse to the king of Morocco and it was to be delivered via an Iranian C-130. They had built a well-padded stall and tied it down in the airplane. The Iranians led the horse up to the plane and started to lead him up the ramp; there he balked and refused to enter.

Being this was the Shah's horse he could not be touched in anger, like hitting him from behind. After about an hour of trying, they decided that it was the stall in the airplane the spooked the horse, so out it came. Still the same old story — that horse would not enter the plane. They were getting desperate; after all, there were great ceremonies planned for the presentation. I was cracking up laughing at their efforts.

Finally I said, "I'll bet I can get that horse into the plane." In desperation I was told to try, but I could not touch the horse in anger. I took a shop rag, covered the horse's eyes, and walked him right into the plane and out again. (I had always been told that you could get almost any animal out of a burning barn by blindfolding it and simply leading it out).

It wound up that I was considered a very "gifted" person, the stall was reinstalled, and I had to walk the horse back into the airplane and into his stall. Then came the big surprise, "Mr. Holdeman, you go to Rabat." The flight went well, and I backed the horse out of his stall and onto the ramp, turned the halter over to his handlers, and stood back and watched all the pomp and ceremony. When it was all over, I got back on the plane and took a return flight to Tehran. The General of the Iranian Air Forces did sent a fax to the Lockheed President thanking him for my assistance.

When I returned to the plant, I was called to the office of none other than Mr. Daniel Jeremiah Haughton, President of Lockheed to explain why that fax had been sent. After I gave my explanation, I thought that Mr. Haughton and a couple of the vice presidents would never stop laughing. After all, Mr. Haughton was raised on a farm. I don't know how much my delivering the horse had to do with it, but Iran became the world's second largest owner of Hercules aircraft. Only the USAF owned more Hercs than Iran.

In the middle of September 1978, I was in Cairo, Egypt when the local office told me that there had been a Herc crash in Tehran and they had requested my assistance. So off I went after visiting the Iranian Embassy, greasing the right palm with $200 for an immediate visa. As always in Iran, I immediately hired a car and driver (we were forbidden to drive in Iran — an Allison representative had been jailed after an auto accident.)

The C-130 accident was fairly straight forward — the right landing gear hung partially extended due to poor maintenance. An inexperienced person had looked through the gear check windows and said it was fully down. The Iranian Air Force General who was flying was late for a meeting and took his word. As the plane touched down, the right side (without benefit of a down and locked right main landing gear) continued to go down until the number four propeller hit the runway and shattered. At that time the General flying decided to go around and applied power. The airplane started a right turn due to asymmetrical power and eventually hit a radar shack and burst into flames. The loan survivor was the inexperienced loadmaster that had reported the gear was down and locked.

The anti-Shah riots were going on at that time, so on the third day there my driver picked me up from the hotel and ran into a mob on our way to the airfield. He couldn't get away, and the mob turned the car over. As I crawled out I was stabbed in the arm but somehow got free and left the area. The Iranian Air Force wanted me to go to the hospital, but I got some butterfly band-aids and used them rather than Iranian hospital stitches.

From then on I had an Iranian helicopter pick me up in the hotel parking lot and fly me to the crash site.

This was the first and only time that I ever saw military personnel deliberately try and misdirect an accident investigation. Some Iranian Air Force maintenance people who were friends of the maintenance people responsible for that plane, removed the left main landing gear retraction assemblies overnight and "modified" their condition at the impact. It was rather crude as the modified position was shiny and the rest of the gear was fire blackened. With that and my wounded arm, I dictated my report and got the local Lockheed office to get me on a Swiss Air flight bound for Athens , Greece which had been my original destination before being interrupted in Cairo to go to Iran. Over the years the scar on my arm has faded. Lockheed was a bit upset and required me to get a tetanus shot and have the arm checked by medical for a couple of weeks.

On June 21, 1977 one of the Navy's secret C-130s was taking off from Wake Island and crashed into the Pacific. The Navy requested assistance so off I went. The full story developed that there was one very hush-hush black box aboard that they wanted recovered at all costs — it was the communication encoding box. About six months before, the U.S. had raised and brought to dock for study a USSR submarine to study. This was a blow to the USSR. Here was their chance to recover. However the United States Navy got a ship anchored over the C-130 impact sight first and hoisted a salvage flag. They then started a mother ship with a DSV (Deep Submergence Vehicle) named the Turtle on board toward the crash site from Hawaii. The Turtle and I arrived at about the same time and after much discussion, it was decided that its crew would dive and try and locate the plane.

After a couple of dives the wreckage was located at 2,800 feet below sea level. Although the Turtle's crew had located the wreckage and had taken some pictures, despite long discussions, they still knew nothing about the C-130 and were at a loss as to how to locate the black box.

Wake Island is really the top of an extinct volcano. During WWII it was occupied by both the Japanese and U.S. forces as an air base. The Turtle's crew and I finally decided that I would dive with them and help them locate the black box since I knew the airplane and could identify areas. The whole side of the volcanic island was littered with both Japanese and US WWII plane crashes, creating much confusion as to which plane wreckage belonged to the C-130. Despite all this, we located the C-130 wreckage, found the black box, recovered it with the Turtle's remote claw, and I saw the underside of a wing observing the flap position. As soon as the black box was in the Navy's hands, they lost all interest in the crash. As they pulled anchor and started steaming away from the site, a USSR ship started coming toward it. The Navy's attitude was "let them have at it" since the black box was safely destroyed — completely melted down; its secrets were safely protected. This system was used for several years after that loss.

As there was no commercial air service to Wake Island, I had arrived on an Air Force C-141 supply flight, so this was the only way back to Hawaii. When the next flight came in, I headed back. I was laying across seats resting and going over the known facts about the crash when I suddenly sat straight up with an idea. The key to this accident was that it was the plane's captain's last flight into Wake Island. He had gone to the chow hall to get some sandwiches for his crew and told them that although it was dark to listen as he was going to give a "low pass" over the chow hall after take-off.

As soon as I got back to Lockheed, I got with a pilot and went through a verbal simulated take-off. Then I went to the Chief Engineer and requested funds to run a computer simulation. He called in the Chief Engineering Test Pilot, who concluded that "no one flying would be that stupid." The Chief Engineer denied my request, but I was so convinced that I had figured out what had really happened not only to this crash but also five others that were not resolved to my 100% satisfaction. All had one thing in common, they had all taken off at night with no horizon and had started take-off flap retraction without having established a positive rate of climb. The resultant loss of lift from flap retraction resulted in impacting the ground/water. With the Chief Engineering Test Pilot being so adamant that "no one was that stupid," I could not legally get the simulation run. So I went to the Computer Simulator Operator who was a friend and asked him to "piggy back" a run with the info that I had given him and let me know what happened. The simulator crashed the plane within 100 yards of where we figured the initial contact with the water had been.

Now I did have a problem, the Chief Engineer had rejected my request, and I had gone behind his back and had the test run without approval. The results were so important that I could only go back to the Chief Engineer with hat in hand, fess up, and take my medicine. Actually, this got his attention and he had a full simulation run made, which actually placed impact within 25 yards of the Unites States Navy's reconstructed initial impact point. That really tore things up. Flight Safety teletypewriter exchange service (TWX) communications went flying out of Lockheed to all operators of C-130 aircraft cautioning against retracting take-off flaps until a positive rate of climb had been confirmed by both pilots and, to the best of my knowledge, is still a caution in the flight handbooks in boldface print to this day.

However, Lockheed and the Air Force were not finished yet. A Flight Safety Engineer and I were flown via a USAF VIP plane to all C-130 bases where they were ordered to hold a safety stand down for our briefing of the six crashes that, by this time, everyone was convinced had resulted from the premature start of flap retraction at take-off. The irony of all this was that at McCord Air Force Base I was asked to sit in on a simulator run. I got in the full motion simulator and went along for the ride. The instructor was having the crew do an engine out instrument landing and then, to add insult to injury, gave then a "go-around" at about 100 feet altitude with full flaps down and configured for landing. The captain added power, initiated a climb and called, "Flaps!" With that the co-pilot brought the flaps full up. The captain yelled, "Freeze!", turned to the co-pilot and yelled, "Didn't you hear that briefing today? You just killed us!"

I sneaked out of that simulator to save further embarrassment to the Air Force personnel. Had I had any doubts at all about resolving the cause of the six different crashes that I had investigated, this erased all doubt. The Chief Engineering Test Pilot didn't have much to do with me for about six months, but I consider this one of, if not the most, valuable contribution I made to the C-130 program. I didn't get a bonus, just a commendation in my personnel folder. I think that it did enhance my standing in the eyes of the Chief Engineer.

Three

What I consider to be my second most valuable contribution to the overall Herc program has to do with engine life. From time to time from various operators we would receive complaints about engine life not living up to the recommended overhaul times. Then one day Lockheed Herc Sales received a phone call from a commercial operator flying Hercs in support of oil well drilling in Mauritania, Africa requesting assistance. It seemed that he was considering purchasing three more Hercs but was having a problem with engine life on this contract. Marketing immediately contacted Engineering and I was assigned to go to Mauritania and see what was going on.

After a few phone calls, I discovered that a commercial customer was operating Hercs out of Nouadhibou, Mauritania in support of Texaco's oil exploration. It took Lockheed Travel by surprise but they finally got me routed to Nouadhibou, Mauritania. The routing came with a hand written-note that said, "Don't ask us what to expect." By this time I was getting a bit suspicious as to what I was getting into. A commercial operator advised that they were operating out of a "camp" that would supply a bed and food.

I went to Madrid then Tenerife in the Canary Islands to await an irregular flight to Nouadhibou, Mauritania. I was met by a mechanic that I had met before at an operators' facility in Miami. He got me to the "camp" and assigned me a cot. They had three Hercs there flying supplies to the oil drilling operation about 1,500 miles out into the Sahara desert.

I had never seen so much sand, both on the ground and flying through the air. I immediately saw why everyone was so upset about engine life. The engines on the Hercs rotated at 13,800 revolutions per minute with close tolerances to compress the air before combustion, which afterwards rotated the turbine, producing the power to fly the aircraft. With the sand ingestion and resultant erosion of the compressor blades and clearances, the compression was down and the resultant power vastly degraded, to the point that getting airborne was not assured due to engine stalls, backfiring, or just plain quitting. In fact, I was on a flight to the drill site and when we started our take-off we had two engines quit. I found myself up over the area microphone to the black box explaining what had happened. I was sure that we were going to crash. The crew reacted perfectly. While flying between two sand ridges they managed to get one engine restarted and we climbed over the sand ridges and into the clear. Later we tried the other engine and it started and we returned to base with all "burning and turning." I think that was the closest I had ever come to believing we were going to crash.

With a lot of hard work, eating a lot of sand, and getting my eye glasses sand blasted to the point I could hardly see (I had to have new ones when I got home) I worked out many procedures both for ground personnel and flight crews to follow and engine life started improving. Under the conditions that the engines were operating, engine life would never reach that of the average Herc, but it did reach "livable" times.

Eventually Lockheed did publish a list of suggestions for the flight and maintenance manuals for operations under hot and or sandy conditions. These suggestions were some of what I had developed in Mauritania, plus inputs from Lockheed contract flight crews operating in several Middle East countries as well as from Lockheed Flight Operations and the Aerodynamics Department. After bringing all these inputs together, they were published as "suggestions for operation in high ambient temperatures and or sandy/dusty environments". There was a lot of "CMA" involved in this as no one wanted to say when "bad was too bad." The basic flight data in the flight handbooks was published and approved based on engines producing their rated power, but no one wanted to provide data for deteriorated power output nor would either Lockheed or Allison establish a minimum power output before repair/overhaul was required. Both contended that this was "operator prerogative." I disagreed at the time and do to this day, but I lost the arguments. However, several operators that operate in such conditions have expressed their thanks to me for the information that had been provided them. I still think that my position was the correct one and believe that some lives have been lost in Hercs because such minimums were not established.

On December 4, 1971 A Navy "Ski Bird" making a jet assisted take-off (JATO) from the Antarctic site known as D-50 approximately 200 miles south of the South Pole had two 165 pound thrust JATO bottles tear loose when they were fired to assist take-off, hitting the number two propeller which fragmented and the flying fragments damaged the number one engine and propeller. The aircraft settled back down and no one was hurt, although the hard landing folded the nose gear. Another "Ski Bird" went in and picked up the crew. Four days later a Navy team came in and in 30 minutes declared the plane a "write off".

Later in 1985 the Navy decided to reevaluate the "write off" decision, and this started a three year effort. On January 8, 1988, Jim Herman, an official inspector from the Navy and I certified the aircraft fit for ferry flight to McMurdo Station for further repair and it eventually returned to full service. I had made one other trip to the crash site in January 1987 to assist in determining what parts would be needed to get the plane ready for ferry flight. I made this trip to help confirm "airworthiness" of the plane for ferry flight after minimum repairs had been made. I then made the decision to fly with the plane from D-50 to McMurdo in case my assistance was needed on any in-flight problems that might occur. Now we were to be escorted by two other "Ski Birds". When Lockheed learned that I had been aboard that ferry flight I was called in and REALLY reamed, and yes, I had known that it was against Lockheed policy to fly on any ferry flight. This airplane had been buried in the snow for seventeen years then recovered and repaired. I would assume that sometime between recovery and now that the plane has been retired — I have been unable to find its history. I do know that new "Ski Birds" had been ordered and manufactured. This was really an outstanding accomplishment. By the time the plane was recovered a new "Ski Bird" was selling for around 28 million and this recover cost a little over five million.

There were many interesting experiences during crash investigations. In Pakistan I had to climb from the 9,000 foot level to the 22,500 foot level of a mountain simply to confirm what I already knew, but the Pakistan Air Force would not accept my statement without "positive evidence." That was quite an effort. It took three days up and two down.

One time a Columbian Air Force plane crashed in a banana plantation on the east side of the Andes mountains in Columbia. From Bogotá the Columbian Air Force flew me to the remote village from where the plane had taken off. I was warned not to drink the water as "it is bad." I never understood why this village had plenty of Cokes but no drinkable water. I was told that the villagers were used to the water and it didn't bother them. I wanted my coffee each morning so that night I boiled up some using my "water heater" and poured it into Coke bottles. The next morning I looked and saw "bugs" swimming around and decided that I could do without my morning coffee.

It was a pretty straight-forward investigation. An inexperienced pilot lost an engine on take-off and mishandled the problem, stalled, and crashed in this banana plantation. It was a gooey mess. At one point I raised up one end of a partial wing just to see a monster snake. I dropped the wing end and ran like hell yelling, "Snake!" The locals went over, picked up the wing part, and grabbed the snake –it was a 20-25 foot Boa; the locals eat them. While at the impact site, which was also close to a rather large stream, a male and female native came up and gathered some plant leaves near the stream, then tool off their clothes, got in the water, and started rubbing each other with the leaves. I was told that the leaves acted like soap. Oh well, back to the investigation.

In October of 1973 the USAF lost a C-130 in the northwest corner of Arizona right at the Missouri state line on a pilot training flight. The weather was sweltering and the investigation took us into the woods in a valley with no wind. That evening, upon returning to the motel, I simply took my shoes off and climbed into the shower. I was soaked from sweat to the point that I squished in my shoes as I walked.

This was a puzzling accident and I was not allowed to take pictures. I had seen the local paper and knew that they had some pictures so went to them and lucked out. I found the person who had taken the pictures and she was the "mother load" of information. She had actually seen the plane at low altitude just before it crashed. The key was when she said, "It was on fire because I saw smoke coming from the end of the left wing" and actually showed me a rather poor picture. As there was not any evidence of an in-flight fire, light bulbs lit up all through my head.

Back at the site I called for the left outer wing fuel tank access plates to be removed. There was the answer — all the internal wing structure was torn loose at the bottom and bent outboard. The airplane had spun, recovered, and when the pilot tried to power up the left engines, tore loose as all their wing attach structure had been destroyed. What the reporter had seen was fuel spewing from the ruptured wing. Some days you just get plain lucky.

However, Military Airlift Command thought that Lockheed was the greatest and that is when the number three item on the C-130 crash check list at Military Airlift Command was changed to, "Call Lockheed and request Ross Holdeman by name for assistance." The Air Force then changed their C-130 flight training syllabus, removing stalling the plane as a part of their training. This decision was made over Lockheed's objection and would later come back to haunt the Air Force over accidents resulting from the pilot not recognizing a stall and taking positive corrective action.

In January of 1981 there was a C-130 crash on the airbase at Ramstein, Germany. I was immediately requested and off I went. There I found that a good friend of mine (affectionately known as Buddha due to his shaved head) was the Air Force Investigating Board President. The site looked to me like a stall with no recovery, but why? This was a supply flight to Norway for the U.S. Embassy in Oslo. The weather was marginal but all proceeded normally until the flight was told to contact Departure Control and expedite take-off since an F-15 was entering final.

Shortly after, the aircraft was quoted as "flying the slowest I ever saw a C-130 fly" by an experienced pilot on the ground. The aircraft had impacted the ground at about 45 degrees, nose down. Investigation revealed that all engines were "burning and turning," the landing gear was retracted as were the flaps, but the throttles were set at a very low power setting. This was confirmed by the propeller blade angles at impact. I was at a loss as to why he was in such a configuration until we discovered his radio was improperly tuned to the Departure Control frequency. I could just see the pilot being reluctant to enter instrument flying conditions with a very high frequency (VHF) radio not working; and with that F-15 on final, he probably wanted to make a wide pattern and return to the base. However, with "all heads down in the cockpit working with the radio" and no one "minding the office" by flying the plane, it stalled at 500 feet and crashed. However, per Air Force procedures, all other possibilities were investigated.

During my second day there the President of the Board came to me and said, "Let's take a walk."

As we walked off he said, "Well, what do you think?"

Being diplomatic I said, "It's too early."

"I know you and you have an idea. What is it?"

I replied, "I am just wondering when in the hell the Air Force is going to teach pilots that regardless of what happens someone has to fly the plane."

He just looked at me and we walked back to the site.

Ross investigates the crash of a C-130 Hercules

I continued the routine and at last made my opinion official to the Board. He told me that they had arranged for a plane loaded as the one in the accident had been and he was going to fly it and try to simulate my scenario. I was already all set to depart so as I left I said, "Watch your ass." He later admitted to me that he almost crashed right on top of the impact site of the crashed plane.

Well, the final report came out as a "Fin Stall" accident. I was upset, but by this time had learned to roll with the punches and refused to change my report to Lockheed. Years later after we both had retired, Louise and I visited him and his wife in Illinois and I asked him what happened. The General of Military Airlift Command had been the head of the C-130 school at Little Rock when they removed stalls from the training syllabus. Enough said. I did report to Lockheed what I had learned, but it sparked little interest.

On September 5, 1980, Kuwait lost a Herc over France. I was dispatched immediately with only the knowledge that it was "in the south of France." Well, Marseilles is in the "south of France" so off I went.

Now talk about luck, upon arrival I went to the Hertz counter to rent a car and find someone who would speak English. While waiting for the paperwork on the car, I looked down at the counter top and under a glass cover was a newspaper clipping in French that had "C-130" and "Kuwait" in English and a date line of "5 Sept., Montelimar, France."

With a lot of arm waving and a Hertz map of France I had gotten them to circle Montelimar and highlight the roads to take. Voilà, I was on my way. I found the closest town, Montel, and a "hotel" and checked in (again with a lot of arm waving) for the night. Then with more arm waving and pointing, off I went toward Montelimar, which I found was a cross roads with three houses, but at one of the roads sat a French gendarme on a small motorcycle. I turned toward him and he held up his hand. I pulled out my Lockheed ID card; he looked at it, held up his hand at me, jumped on his motorcycle, and off he went. In a few minutes he was back motioning me to follow.

We topped a slight rise and there was the impact site. I got out and a Frenchman came running up to me and said, "I call Paris one hour ago and, voilà, you are here." (He had called Paris and asked them to send Lockheed assistance).

The crash site was a big hole in the ground — the most complete destruction of a C-130 that I had seen. There were some bits and pieces that I recognized. Then with the help of a French to English and English to French dictionary, we did work out that there was another part over the mountain. As it was late, it was decided to make the hike the next day.

So bright and early the next morning off we went on our hike. It was fairly rough going but we found a large portion of the left wing. At some point, I don't remember my source, I learned that the French air traffic control had picked up a frantic call from the plane saying, "We are broken. We are broken," several times. There was no evidence of a fire and I took a closer look at the left outer wing.

Suddenly, I noticed that the aileron moved freely on its hinges. Further investigation revealed that a bolt was missing from the aileron push pull rod to the bell crank, thus allowing the Aileron to float freely. Problem solved. That bolt was supposed to be installed with the head up so that if by some quirk the nut came off the bolt would remain, attaching the push pull rod to the aileron bell crank. The ailerons on the Herc did not have any aerodynamic mass balances to protect against flutter as the push pull rods acted as mass balances. It was known that if the ailerons were floating free, they would flutter and excite the wing into a flutter state that would result in wing separation.

Later that day, after we returned to the impact site, all hell broke loose — one of the people had dug up a mass balance from the rudder. To save space but get the required weight Lockheed used depleted Uranium, which is heavier than lead for the same mass. However, by law, these mass balances were required to have a decal with the atomic symbol installed on then. Even though the identifier contained the word "depleted," all hell broke loose.

I was forced aboard a French van with the French gendarmes and that mass balance and off we went tearing along roads with sirens screaming. I didn't know where we were going, but just for the hell of it I sat down on that mass balance.

After a wild 40 minute ride we came to a halt, marched into a large guard house, were photographed, issued badges, and went in. In a lab someone finally spoke excellent English and understood what I had been saying all along. A lab technician confirmed there was no radiation and with a lot of arm waving and French jabber we finally all loaded up in the van and went back to the crash site.

I later learned that we had been to the French National Atomic Energy Laboratories. The French informed the Kuwaitis of their (my) findings and all hell broke loose — this could not be a maintenance error. Kuwait threatened to cut off oil to France unless the French made their findings "more realistic." The French report was then changed saying that a "lightning strike" resulted in an internal wing explosion that caused the crash. Case closed, well, almost.

About a year later the French courts, they handle all accident investigations, requested I come to Paris for "consultation." We thought this over and decided that the Lockheed Chief Test Pilot would accompany me, just in case. Anyway, we left a couple of days early and I took him to the French Air Base where the remains were still stored and then on to the impact site and offered to take him over the mountain to where the wing was found, but he declined. We satisfied the French courts with their questions and returned home. The Chief Test Pilot expounded considerably as to what I went through on that accident investigation, and it was actually an easy one.

An interesting crash occurred October 1, 1979. Early that morning, well before daybreak, a Bolivian Herc was taking off from Panama International Airport. They were taking off into a "no horizon" sky over the Bay of Panama. They impacted the bay.

The tides there run to twenty feet and the bay is full of silt. When I arrived, all that could be seen was about the upper half of the vertical stabilizer.

As this was the weekend that President Jimmy Carter was turning the Panama Canal over to the Panamanians, there were only third-rate hotel rooms available. Anyway, I struggled a bit but finally found a father and son up-river that would take me out to the site, for a price. However, they warned me that timing was everything as tides ran about twenty feet. So at the agreed time, I arrived to find a real honest-to-goodness dugout boat and a thunderstorm in progress. We were delayed about twenty minutes waiting for the storm to pass and then started down the river in the dugout to the Bay of Panama.

The delay resulted in the dugout bottoming out on the mud about 200 feet from the site. I got out and immediately sank up to my waist before I could climb back in the dugout. Well, we were not going anywhere until the tide came back in. I thought about this a bit, then I climbed out of the dugout and found that by laying flat on my belly I could stay on top of the mud. I started moving by fully extending my arms and paddling with my hands. I could "slither" along this way and made it to the impact site. There I was able to check the trim tab positions, which were all normal, and was actually able to determine that at least two engines were developing take-off power. Then I lucked out and found a wing portion with a flap jack screw attached and found that the flaps were about 75% retracted. A bit later, I found three other flap jack screws and confirmed the flap position. By this time I was getting pretty worn out from "mud swimming" and found a large section of a wing plank just under the surface and got on it to rest. About that time I noted that the tide was coming in and started yelling and waving my arms to the dugout to come and get me. They had just started to float again and in they came. I had my answer, premature flap retraction, but boy was I a mess. Fortunately, I had bought a cheap pair of coveralls and put them on before I started the dugout trip. We got back to the river and their "landing". I got out of the dugout into the fresh water and washed off a bit and then changed into my "work clothes."

During this time, while paying off the locals, it was casually mentioned that the City of Panama's sewage was dumped into that bay. This scared me as my hands had gotten cuts and scrapes as I "crabbed" about from all the broken aluminum. It had always been my policy to buy a quart of Canadian Club at the duty free store when leaving the country. So when I returned to my hotel room I opened that bottle, took a drink, cried a bit over what I was about to do, put the stopper in the wash basin, dumped the bottle in it and washed and washed and washed my hands. I didn't get an infection.

When I was debriefing to Lockheed after they got through giving me "what for" for such a stunt, I informed them that I expected them to approve the purchase of that quart of Canadian Club and furthermore, they had sent me to Panama to find out why that plane crashed and I had done just that. At the end I was politely requested not to list the Canadian Club on the expense account. When I turned in my expense account I attached a note saying, "It's in here, just find it." The expense account was approved and a quart of Canadian Club found its way into my car at the plant with no note attached.

Another interesting incident occurred when I was sent to Vietnam to assist in an operational problem encountered by the C-130A gunships. You really have to see what is going on, verbalizing just isn't sufficient sometimes. So, after signing all the liability wavers, I went on a C-130A gunship mission over the Ho Mien Trail to "shoot up a supply convoy" coming out of North Vietnam going to their people in South Vietnam (again Lockheed people do not go into combat situations). While we were flying over the Ho Mien Trail destroying the supply convoy, we were surprised by intense ground rocket and small arms fire. Before we could extract the plane and ourselves, the number three engine was hit, blowing the propeller and front half of the engine gear box off. They crashed into the number four engine knocking it askew and the entire engine gear box and propeller off. Now we had only two engines on the left wing working, which was bad enough, when a surface-to-air missal glanced off the aft fuselage, fortunately not exploding, but tearing out a large chunk of the empennage (tail section). Now we had several problems with only two engines and a heavy airplane at low altitude.

The pilots did a fantastic job maintaining semi-control and keeping the plane in the air while all hands started throwing everything out of the plane that was not nailed down to reduce weight. With flight now stabilized, I noted that, due to the hole in the empennage with the asymmetric thrust being countered by full left rudder, the empennage was wrinkling badly. So with me on a headset in the back of the plane watching the tail twisting and talking the pilots into a balancing act between keeping the plane in the air and not so much asymmetric power as to twist the tail off we managed to get the plane off the trail and back to base.

In the end the plane was written off, the crew was given metals, and I came back home. For some reason that I never understood, I was allowed to take pictures of that plane after we landed, I still have some, and when I showed them at debriefing — talk about catching hell. My only defense was, "You sent me to do a job, I did it, and you got paid". I applied for the oversea hazardous duty bonus and although I was not there long enough to qualify for it and it was approved. I don't think that Louise ever knew exactly what the amount of my check was until she got the bank deposit slip after it was directly deposited.

Speaking of gunships, I think that it was in the fall of 1968 that the C-130 project was being moved to a new location and we were getting all new desks, etc. Therefore it was a "box move." The layouts were all made and Monday morning we were to report to our new location with our boxes being at our assigned desks. We were free to leave that Friday when we were all packed. I got home around 4:30 as I recall. About 5:30 the phone rang. It was the project Engineer. He said there was a C-130A at Eglin Air Force Base Florida that could not get the number three engine started, and there was a Lockheed Electronics Company representative there and I should call him at such and such a number. I asked if I was authorized to go if needed to and he said no.

I was talking to the representative when the phone operator interrupted and said that I had an emergency call. It was the Chief

Engineer again. He told me he had misunderstood and there was a Southern Airlines flight leaving Atlanta in about two hours for Eglin and I was to be on it and would be met upon arrival.

Well, it had been about three years since I had worked on a C-130A aircraft and all my "brain books" were packed up since we had just moved. When I got to Eglin and was getting off the plane there was an Air Force car, two one star generals, a lieutenant colonel and a staff sergeant there to meet me. The sergeant asked if I was Mr. Holdeman and I said yes. He asked to please give him my baggage check, he gave it to an airline employee, they got my bag, and the sergeant led me to a staff car. We drove across the field to a C-130A.

On the way over I was briefed that the airplane had been extensively modified and the Secretary of the Air Force was to arrive Sunday for a night demonstration of the plane's ability. They were unable to get the number three engine started and it was absolutely imperative that I get things in working order. The nearest C-130A engine mechanic was in Vietnam and they could not get him here in time. I told them that I had Top Secret clearance but I had no proof since clearance usually had to be transmitted from Lockheed to the base Security Officer and Lockheed was shut down for the weekend. It didn't faze them a bit. They told me, "Just get that engine working."

I was getting a lot of "help" from the four generals and they were getting to me. I walked away to think and the lieutenant colonel came up and asked, " Are they getting in your hair?"

I said, "Yes."

He turned and said, "Gentlemen, we are going to the motel. Mr. Holdeman, the sergeant (who was the flight engineer) has the number if you need anything or get things working."

Talk about lucky — at about 5:00 A.M. I found a relay that was not working. The sergeant and I found a similar one in base supply and lashed up a bracket, hooked it up, and the engine started. We then had to fabricate a mounting bracket and make the installation air worthy. This done, he called the lieutenant colonel. He came out and actually took me to a motel to get some sleep saying he would pick me up about noon.

He did and I asked to be taken to the civilian terminal to see about a flight home. He said no, I was staying until the Air Force Secretary's flight was over Sunday night and they would give me a contract. That was something new so I called the Lockheed Chief Engineer. He told me to reject getting a contract but to stay as part of "Lockheed's good support policy." I got a rental car and a motel room and settled down. A knock at the door was the sergeant who said they were going for a test flight and the lieutenant colonel (who by this time I had concluded was "Big Daddy") wanted me to come along. OK, so back to the field.

The flight really was uneventful, although they tested all the special equipment (boy was I amazed at what I saw) and were satisfied with the plane and its equipment. When we landed the flight engineer sergeant was called to report to some office on base. There he found that his wife and children had been in a very bad car crash in Ohio and he was given emergency leave to depart. Everyone thought that they could get a flight engineer from the C-130 unit at Eglin. Then they found that there were no C-130A flight engineers there nor could they locate one in the States. I was asked if I could work the panel, and I said I could. That settled it, I would be the flight engineer on the big flight. I tried to get some official at Lockheed to give me guidance. I could not reach anyone. I then got to thinking, in an emergency would I react correctly and suggested that they get me a C-130H flight engineer, give me several hours to check him out on the C-130A panel and an Air Force flight engineer could make the flight. This was agreed and I spent most of Sunday "checking out" a C-130H flight engineer on the C-130A panel.

At the last minute the lieutenant colonel came to me and said, "You will go along on the flight as a civilian technician." I was introduced to the Secretary of the Air Force as such and kept a watchful eye on the flight engineer during the flight. All went well. I was absolutely shocked at what I saw on that flight. After the flight, I was told I could go home on the first available flight. I got into the plant about noon on Monday, and as soon as I found the C-130 Engineering Group in their new location, I was grabbed and taken to the round "Secure Room" where I was debriefed and read the riot act about not telling anyone outside the room what I did or saw. There was no overtime and no bottle of booze in my car after this trip.

Anyway, this was the prototype of the AC-130 gunship. Subsequent to this there were 18 C-130A aircraft converted to the AC-130A (gunship configuration). Over the years there have been many configurations and they have been utilized in every combat area, and a few undeclared combat situations, since that time. In fact, there is a squadron of black painted AC-130 aircraft based at Kendall Air Force Base Florida, about 50 miles east of Spanish Cove; sometimes you can see one taking off from the US 98 highway. Lockheed Georgia has never been involved with the conversions or configurations of these planes. However, I believe that Lockheed Air Service Company has done at least repair/modification work on some as I had a call or two before and since retirement. This use of the C-130 as a gunship has been so successful that, to the best of my knowledge, there have been about 50 C-130s modified to various gunship configurations since the initial 18 C-130A modifications.

Four

There was interesting occurrence in May of 1979. Lockheed received a very long TWX signed "Mr. Lacomb, Director of Maintenance for Angola Airlines" (TAAG) requesting assistance in an accident investigation of their Lockheed Hercules aircraft. Well, first of all, Lockheed did not know that Angola had a Herc. So back went a TWX requesting the serial number of the airplane. Back came a TWX identifying the aircraft as Lockheed Serial Number 4176. A bit of checking and phoning revealed that Lockheed had sold that plane to Delta Air Lines who had sold it to Alaska Air Lines who had sold it to Angola Air Charter, a branch of TAAG.

    We quickly learned that Angola, although not formally embargoed, was not considered a "friend of the U.S." Although Cuba had several thousand "technical advisers" in Angola, there were no travel restrictions, however, there was also no U.S. Embassy. I was sent to Lisbon, Portugal where it was thought that I would get a visa at the Angolan Embassy there. The Director of Maintenance for TAAG (who had requested assistance) was sent a TWX that I was en route to Lisbon and to advise their embassy I would be there for a visa.

When I arrived in Lisbon and went to the Angolan Embassy they had never heard of me or of Mr. Lacomb and were not impressed with the TWX that I showed them. The best I could get was a visa application. I went to the U.S. Embassy and as usual, they were of no help. I decided to go on as I had a confirmed flight reservation with TAAG and would try to bluff it out. After all, the Director of Maintenance had asked for Lockheed's help.

I got to Luanda about 10:00 P.M. and was barred from leaving the airport. I had resigned myself to sleeping on a bench when I saw a station wagon drive by with the letters "TAAG" on its side. I yelled like mad; the driver stopped, backed up, and came in past security. I showed him the TWX and fortunately he spoke English. By leaving my passport with emigration he got me away from the airport and took me to a place where Mr. Lacomb was but would not let me go in. The driver went in and came back telling me that he was to take me to a hotel. What a hotel — it was obvious that at some time in the past, probably when Angola was in Portugal's possession, it had been a very good hotel but now! I was given a room on the 6th floor and the elevators didn't work. My room did have a bed and a sort-of bath but no soap. Anyway, at least I had a bed.

The next morning I walked back down to the lobby and finally got them to call Mr. Lacomb. After several attempts to get the phone to work, they reached him and gave me the phone. By this time I was a bit upset, but Mr. Lacomb said that a Russian Izu car would come and get me. I was really getting steamed but kept my composure. Finally, I got back to the airport and the TAAG offices and Mr. Lacomb. I expressed my displeasure with the way I had been treated and insisted that he retrieve my passport with a valid visa.

Then I learned the favorite Angolan expression, "no sweet". He picked up his phone, someone entered his office, and after a short discussion in Portuguese, off they went and returned in about five minutes with my passport with a stamped visa.

It was then that I found out that the accident was not in Angola but on an island named Sao Tome located on the Equator about 200 miles east of the Eastern Coast of Africa. Mr. Lacomb informed me that he would have a TAAG plane fly me and the Lloyds of London Insurance Claims Adjuster there .He further informed me that there were no communication facilities on the island.

I agreed to go and wrote a TWX to Lockheed that I would be out of communication for a few days. I later found out that this caused a great uproar at the plant with them contacting the State Department.

So the insurance adjuster, a "Hercules mechanic" that "serviced the plane" ,and I set off for Sao Tome in our private Soviet IL62 passenger aircraft complete with Soviet pilots but an Angolan cabin crew. Breakfast was served and after about 2½ hours we landed in Sao Tome.

We were greeted by the local TAAG agent with an English to Portuguese to English dictionary in his hand and an immigration official in tow to stamp passports. About that time I heard jet engines and saw our plane taxing out and taking off. Well, that Lloyds of London adjuster stuck like a leach to me. With the help of the dictionary and the "Hercules mechanic" we understood when the local TAAG agent told us that he had "rooms for us at the number one luxury hotel on the island" and a car and driver, but he would be there to assist. Our bags were placed in the car and we were driven out to the accident.

The Herc's tail was about 50 feet clear of the runway. I immediately noticed that the number four propeller looked to be feathered and the number three almost feathered. I also noticed that the right landing gear was folded under the plane and the right wing tip was on the ground. There was a fuel leak from the number four wing's fuel tank and the vertical tail was askew by about twenty degrees with the resultant fuselage wrinkles. It was a pretty sad situation for an airplane to be in.

Eventually, with the considerable aid of the English to Portuguese to English dictionary and from what Mr. Lacomb had told me, it seemed that TAAG had bought the Herc from Alaska Air Lines and they had checked out the pilot in the airplane (probably about a 30 minute effort). I also learned that the pilot was "very experienced" as he had been flying old DC3 planes for many years.

It developed that the plane had been sent to Sao Tome with a load of freight. When the pilot took off for the return trip at about 10,000 feet the number three engine and the propeller's RPMs started fluctuating. As the weather radar was inoperative and there was a thunderstorm in the area, the pilot decided to return to San Tome. At touchdown the brakes failed; the pilot couldn't stop the plane and as the runway ended over a 400 foot cliff, he decided to ground loop the plane — at least that was his story. From there on I can only reconstruct what happened based on the physical evidence. The pilot must have been in a panic because he dove at the runway, touching down about 2,000 feet down the runway with first one main landing gear and then the other as he was too fast for the plane to stay on the ground. The brakes didn't work because at that speed the plane never got heavy enough to make the touchdown switches, thus the normal brakes would not engage and the emergency brakes were never selected. The crash was due to pilot error all the way! Now I had a problem since the pilot was considered to be TAAG's best. I decided to be diplomatic and write a report simply using the cockpit switch positions and the physical condition of the plane.

After three days there I got the local TAAG agent to radio TAAG in Luanda that I was ready for pickup. Word came back that a plane would be down the next day. The plane with the same Russian flight crew arrived and I expected to head back, but no — first the plane pilot and co-pilot (who spoke respectable English) wanted to see the crash site. So out we went back to the end of the runway and the crash. They did a respectable job of looking the plane over. After this they wanted a beer so went into town to a beer garden where we met a Russian sea captain who had been studying English and wanted a "real American" to practice on. To make a long story short, that is how I found out that the DC-10 was grounded by the sea captain asking, "Why 10 DC no fly?"

Of course I had no idea why, and in those days that was about the only plane being used between Portugal and the U.S. Well, we finally headed back and were all good buddies by then. The crew gave me Olympic coins as that was the year the Olympics were held in Moscow and Jimmy Carter barred our participation.

On the way back I was invited to the flight deck and allowed to fly the plane — it was quite a thrill. Upon return to Luanda, I was sent back to the same hotel and my six flights of steps to walk up.

The next day one of the twice-a-week flights back to Lisbon was scheduled to arrive. I insisted that they get me on that flight. But first we had to have a "debriefing" meeting. All went well until we got to the propeller blade positions and I was asked how they could get that way.

The pilot and co-pilot were in the meeting, and I said that since the condition levers were in the feather position, I assumed that was the reason — I was still trying to be diplomatic. A big discussion ensued with the co-pilot finally saying, " Maybe he did it but did not remember doing it.

I looked at the clock. It was past the Lisbon flight's departure time and I heard the engines of the DC8 running, but Mr. Lacomb kept the meeting in session, saying that the plane would not leave until he okayed it. I did a lot of squirming but finally was taken to the plane. It took off for Lisbon with me aboard, breathing a sigh of relief.

In Lisbon I bought a Newsweek magazine in Portuguese (no English copies were to be had) and saw the pictures of the Northwest DC-10 Chicago crash. I went to my favorite hotel in Lisbon, called Lockheed, and got them to get me scheduled for flights back home.

That evening at dinner I had company. It was an embassy official who tried to give me the third degree. I really had a bit of fun evading his questions, but the next day on the flight from Lisbon to Frankfurt I had a seat mate that kept trying to pump me about Sao Tome, Angola, and the South African Air Force (SAAF). As I evaded his questions and ignored him, I placed him as a U.S. Embassy spook and discussed my displeasure with the embassy when I needed a visa to Angola. I flew from Frankfurt to Atlanta on good-old Delta.

As soon as I hit the plant it was off to the round "Secure Room" for about five hours of debriefing. I got all over the CIA people about the Lisbon Embassy's people. They had already gotten a report and actually congratulated me on how I had handled a situation that they didn't know was happening.

A few interesting footnotes to this was that while in Sao Tome I had gone to the hospital to visit the loadmaster who, while the airplane was sliding sideways down the runway, had decided to jump out of a paratrooper door. In doing so he got his foot caught and his head hit the runway, bouncing along, and the whole left side of his head and face was scraped off. I gave his wife some money to assist with expenses while she was in Sao Tome and my Lockheed business card. About two months after I returned, I was called over to Legal and shown a "Notice of Suit" document against me by name. They wanted to know what I knew about it, so I told them of my visit to the hospital and how I gave the couple my business card. So between us we finally reconstructed that the loadmaster had gotten a lawyer and as he had my card, they filed against me. I asked, "What do I do?" and was told, "Nothing. We will take care of it." End of story; I never heard anything else and never asked any questions.

The second interesting sideline to this was that at the Lockheed debriefing I stated that we probably could sell them some Hercs. The initial reaction was, "They are embargoed by the State Department." To which I replied that Boeing had a five member team down there working out the final details for a purchase of twelve Boeing 737 aircraft, and if they could sell there, why couldn't we sell the civilian version of the Herc. After a few moments of silence the Vice President of Marketing spoke up and said, "I'll get someone down there." Within three months Lockheed had a firm order for seventeen civilian-version Hercs complete with State Department export approval. There was, however, no sales commission for me.

Another interesting sidelight to this trip is that when I had first gotten to Sao Tome and was on the way to my hotel, I noticed all the groves of trees. It looked to me like they were cultivated, but I could not place them. After much trying, I discovered that they were cocoa bean trees and that the fruit on them were the beans that cocoa was made from. I had never seen such before. Anyway, I finally got it across that I would like to buy a couple of those cocoa beans. So at some point the driver that took me to and from the hotel stopped, jumped out of the car, ran over to a tree, picked two of the cocoa beans, ran back to the car, and handed them to me. I really had wanted them to show back home and then give them to my grandson. Well, when I got back home to Atlanta and came to Customs I opened my brief case and there were the two cocoa beans. The Customs Agent asked about them and I told him that they were cocoa beans.

He looked sort of funny at me, handled them, and yelled to the Agriculture Inspector, "Hey! Are you interested in cocoa beans?"

The reply was, "What?"

The Customs Agent replied, "Cocoa beans!" and with that the Agriculture Inspector walked over, looked at them, and I explained where I had gotten them and why I wanted them.

He handled them and said, "Well I'll be damned. I never saw one before and don't blame you for wanting to give it to your grandson" and walked away. My grandson got the beans after I had shown them around home a bit.

I had one more encounter with Angola. In February of 1992, after I had retired and been in the Hercules consulting business, I had a call from a Mr. Gregg Kielton, President of Aircraft Service Group in Miami requesting a copy of our rates and personal information. I sent this and later was requested to come to Miami for a meeting on May 10th. After confirming that our out-of-pocket expenses and my regular time rates would be honored, I flew to Miami on May 9th for a meeting the next day.

I arrived at the Aircraft Service Group's office and was introduced to a Mr. Erich Kouch of Transafrik Corp. LTD located in Luanda, Angola and with that Mr. Kielton said, "Use my office, I have something else to do."

It seemed that the Transafrik Corp. was operating a Herc that TAAG had leased from South Africa Airways and was under contract to TAAG flying supplies into the interior of Angola. The contract was coming up for renewal and TAAG was accusing Transafrik of improper maintenance and unsafe operations. Transafrik thought that this was a tactic to try and get a reduction in the contract price. My job was to inspect the airplane and Transafrik's operation of it and provide a report. As I had no great love for Angola, I stated that I would quote a "package deal" for five days of on-site work. Transafrik would provide in country transportation and first class accommodations. I was told that there were no in country first class accommodations but they had a company village and I would stay and eat there at no cost. I would inspect the aircraft records and accomplish physical aircraft inspections between flights with them supplying the mechanics to open all panels I directed, etc. Furthermore, I would draw up a contract and fax it to Miami for approval or rejection. I expected that the total package would run approximately $40,000, but after I worked out the inspection requirements and determined airfares, I would fax a final figure and if they accepted in writing, I would start work.

With that I presented my bill for the trip to Miami and Mr. Kouch went to the secretary of Aircraft Service Group and asked her to issue me a check for my stated amount. I returned home and went to Lockheed to confirm that I was not "stepping on their toes" on a job that they would like to have. There I found out that Lockheed was the source of my being recommended to Transaferik for the work. The excuse I got from Lockheed for there not doing the job was that it was too small — "Our paper work would run more than the total price you are thinking about and we know that you will do as well a job as us."

In about ten days I faxed a copy of my proposed inspection plan and a firm quote of just over $38,000 for five days on site and $5,000 a day if I was required longer, and they would have to secure my visa. I wasn't sure whether I had over priced or not (actually there was a "negotiation factor" in the quote). In a few days I received a fax requesting an arrival date in Luanda. This was negotiated and soon I was off for Angola again.

I made it to Lisbon and there things sort of fell apart. The TAAG flight I was scheduled on broke and I had to scramble and get a room in Lisbon for the night. However the next day the TAAG flight went and I arrived in Luanda one more time with no visa, but this time I was escorted off the plane directly to an Angolan official who took my passport and stamped it on site. To make a long story short, at their request, I stayed an extra day for a meeting with the TAAG people assuring them that the plane was being maintained in proper condition and I had even flown the plane to prove to myself that all was well. This was the last time that I flew a C-130. I guess it all paid off because Transafrik got their contract renewed for two more years, however, it would not be completed as the civil war got to the plane with a ground-to-air missile and it ceased to exist about nine mounts after I was there. I found this out from Lockheed with the added comment, "Now you see why we didn't want that contract". My reply was, "With friends like you guys, who needs any enemies?"

The only down side to this sojourn was with my extra day in Angola and the lack of communication facilities, I could not get word to Louise and with time changes I didn't want to wake her up at 3 A.M. with a phone call from Lisbon.

With a late arrival in Frankfurt I had to run to catch my flight to Atlanta. At the Atlanta Airport I grabbed a porter but there was no time for phones as he was off to my car with the bags. Anyway, when I got home there Louise was nowhere to be found! I almost panicked, as tired as I was, but then I thought that she must be at a friend's. I went over and, sure enough, there she was and in front of everyone I was chastised for not calling. I couldn't really blame her.

One of my fondest memories started when I was in Edmonton Canada with Pacific Western and Louise called and said that my granddaughter, Tracy, who was in the 3rd or 4th grade, wanted me to call her. I did and Tracy asked me to tell her all about Eskimo life as she had a report to do on that subject. I told her that I was flying to Resolute Bay that evening, which was at the magnetic North Pole, and there was an Eskimo village there and I would try and get some information and call her back by Sunday from Edmonton.

I flew with Pacific Western from Edmonton to Resolute Bay and fixed the problem with one of their Hercs before it went into deep freeze at 48 degrees below zero. I then borrowed a Snow Cat (an enclosed tracked vehicle for traveling over snow) and drove down to the Eskimo village and found the "Head Man." He was exceptionally nice and I told him what I wanted. He spent a couple of hours telling me about their life while I took notes. Then I traveled back to the air base and finally back to Edmonton. I called Tracy and read her my notes.

Later I was told that she got an A+ and her teacher asked if she had just called her grandfather." Tracy said, "Yes" and told her about my trip to the village. For some time I had been writing Tracy (along with Louise and her mother and my mother and others) a picture postcard from every country I visited with a bit of interesting local information. Then Tracy asked me to also write one to her 4th grade class. This added to my postcard writing and before I knew it I was writing and mailing 26 post cards from each country I visited. That was a lot of writing, however most of them were really duplicates of the information, just sent to different people, but it still had to be written by hand. Sometimes it got rather expensive — I paid as much as 50 cents for each card and 67 cents postage. Although it was a lot of work, it made the evenings go by and I preferred writing over sitting in the hotel bar. I received lots of compliments from the recipients of the cards, so I know my efforts were appreciated. I thought the stamps would be interesting to a collector, but the people that I talked with about the cards just kept them.

Then one evening in Athens, Greece, after a good dinner and plenty of wine, I was walking back to my hotel and passed an open shop displaying painted miniature replicas of ancient water jugs, pitchers, bowls, etc. I went in and started negotiating for 30 of a mixture of these miniatures. It was in November and I was headed home in a couple of days. I knew our plans were to spend a few days at our daughter's house. Tracy had already promised her class that I would come have lunch with them.

As Louise and I were going to have lunch with Tracy's 4th grade class, we went over a bit early to hand out one of the miniatures that I had brought to each member of her class. We went into the school office to ask directions. I identified myself as Tracy's grandfather and that started a bit of an uproar in the office; it seemed that when a card from me arrived addressed to "Miss Page's 4th grade class," it immediately started making the rounds among all the teachers and administrators before winding up in Tracy's class. Louise and I finally made our way to the classroom and explained what I had brought. There was great excitement and it was decided that to distribute the miniatures, the students would file by and I would reach in the bag without looking, grab a miniature, and give it to the child. Thus there were no favorites but the children could trade among themselves if they so desired. I was amazed to see a world map on the wall surrounded by my post cards with colored yarn leading from the card to the country from where I had sent it. I thought that was really great.

Anyway, after things settled back down, Miss Page asked if I would answer questions. I was getting concerned about interrupting the class routine but answered, "Yes". I will never forget the first question; it came from a little boy and was, "How can you afford to go to all those places?" I explained that it was my job and a big company paid me to do it. There were quite a few questions but then it was lunch time and Louise and I ate with the class. Afterwards we departed. But those miniatures started a new tradition— in each country I visited I tried to find 34 "somethings" and send them to the class. I sent rubber tree seeds, Arabic Easter cards, miniature Leaning Tower of Pisas, etc.

One other interesting thing that I did for my granddaughter was buy a doll in every country I visited representative of the people's look and dress. I think that Tracy finally accumulated 62 dolls. One Christmas Louise and I gave her a stand-up glass display case for them. I think that at various times some or all of them were displayed at her school and various other places. It was a lot of work but enjoyable and as Louise says, "kept me off the streets and out of the bars."

The Hercules was a good subject for pictures and I always enjoyed photographing it. Some of my best Herc photos were from the far North above the Arctic Circle. I was with Pacific Western one time on a Canadian government chartered flight to get supplies to several remote Eskimo villages. I thought it was great to see the supplies the plane was carrying loaded onto dog sleds for moving them from the frozen lake where we had landed over to the village. Those pictures were very popular back at Lockheed. One of the pictures I shot on the North Slope was of a Herc landing at "high noon" with the sun just barely above the horizon. That photo and the Herc both looked great on the cover of the Lockheed magazine.

Pacific Western was the first civil operator that operated the Hercs continuously in sub-zero temperatures. Although the military had operated the "Ski Birds" in both the Arctic and Antarctic, they had a considerable amount of ground support equipment and making a profit was not their worry. Civilian operators utilized the "vanilla Herc" with very little special ground support equipment and many problems occurred. As a result I really got a workout in civilian operation of the Herc in sub-zero temperatures.

Therefore, when the North Slope oil boom came along and oil drilling equipment was built to "fit into Hercs with ½" clearance side to side, top to bottom and fore and aft" I became a regular face from Fairbanks to Prudhoe Bay. In fact, at one time I made three trips from Lockheed to Fairbanks in two weeks. Every civilian Herc that could be contracted at outlandish prices was flying from Anchorage or Fairbanks to numerous ice strips from the north side of the Brooks Range to the Arctic Ocean. The inevitable occurred — Herc flight crews and maintenance people from Florida and other southern states were dumped into 30-60 degrees below zero temperatures with white outs and as much as 21 hours of darkness — it didn't work out so well. Accidents and crashes occurred but the lure of big bucks kept them there and me "commuting" form Atlanta to Fairbanks solving mechanical and operational problems.

On one trip I needed a part for a plane and knew that it was probably undamaged on a plane that had crashed on final approach to Put River. So I went on a flight to Put River and went to the plane to remove the part. All was well until I started to crawl out and heard something. Using my flashlight I looked around and found a fox barring my way out. Most wild animals on the North Slope were rabid so I had a problem. He didn't seem afraid of me and I was not dressed for extended exposure to -45 degree temperatures. I finally got a long piece of the floor and jabbing at the fox, backed him away so I could crawl out of the plane. By this time I was so cold that I literally had to keep telling myself out loud to pick up one foot and move it forward and then the other until I got back to the oil rig shack. That was a close one.

At one point Lockheed "ran by me" the idea of having me set up a maintenance group in Fairbanks; the problem was that due to insurance liability, it would have to be completely separate from Lockheed. Although the terms were most attractive, I passed and kept commuting between Atlanta and Fairbanks during the winters. Things sort of came to a halt in the summer as it was a lot cheaper to move supplies by barge than air. It really was not until I retired and Louise and I took a motor home trip to Fairbanks that I saw the ground there, before that it was always covered in snow and ice — it sure looked different. In fact, I had a bit of difficulty finding my way around Fairbanks. We had our two dogs with us, so while I stayed in Fairbanks and "doggie sat," Louise went on an group overnight flight tour of Nome and Prudhoe Bay. I was so glad that she got to see that operation.

One of the North Slope crashes occurred right on the Arctic Circle at an oil camp support base named "Old Man Camp." I was in Israel assisting in the recovery of a damaged Herc out of the Sinai Desert. The plant sent two totally inexperienced people to support the investigation in the Arctic Circle. They came home in four days saying a vertical gust had overloaded a wing and it broke. The National Transportation Safety Board (NTSB) investigator brought back some wing structure for the NTSB lab in Washington to "analyze for fatigue." I got home Friday evening and Saturday morning the Chief Engineer called the house to see if I was home. He got me on the phone and requested that I come to the plant and listen to a copy of the cockpit voice recorder tape. So out I went and wound up shocked. After listening to the final section several times, it became obvious to me that the number three engine had somehow gotten loose and hit the right side of the fuselage right at the cockpit; you could actually hear the cockpit's window's glass breaking . A call to the NTSB in Washington revealed that they were not satisfied and wanted a Monday meeting. So off I went to Washington Monday morning and Monday evening was on a flight to Fairbanks with the NTSB investigator.

I had an idea of what had happened but would not reveal it until I had more facts. To make a long story short, we went back to the impact site from Old Man Camp (about three miles) via chopper, hovering on the side of the slope, while we unloaded. A caterpillar had started out first pulling a snow sled to recover the number three engine which was approximately one quarter mile from the main impact site. Some parts of the number three nacelle were never located — that helped confirm my suspicions. Anyway, we finally got that number three engine back to a hanger in Fairbanks and after about 30 hours in a heated hanger, the engine was thawed enough to start an inspection.

Probably three to four hours into our inspection of the engine and nacelle I suddenly stopped, called the NTSB inspector over, and pointed to the rear engine mount and structure. Voilà, there was the cause of the crash: improper securing of the engine's rear engine mount to the nacelle structure. This had allowed the engine to start swinging around the front mounts and enter into what is known as "whirl mode," which is where the propeller starts wobbling instead of being restrained in its normal plane. The abnormal forces generated tore the engine and propeller loose from the wing, taking the wing structure, which hit the side of the cockpit, resulting in the wing separating from the airplane.

I had rented a car and the NTSB investigator was riding with me. We took that part of the nacelle and put it in the trunk of my car and I handed him the keys. I said, "You know that there are going to be law suits over this. You keep the keys except when I am driving and we both are in the car. I never want access to that nacelle without you present. I can't let you drive due to Lockheed liability."

He finally had the nacelle structure crated and air shipped to the NTSB lab in Washington. The engine went "in bond" to the Air Force Plant representative at Allison in Indianapolis for secure tear down. This was accomplished under the eyes of the NTSB with me on the sidelines.

During tear down, the condition of the rotating parts of the engine were exposed and all doubts regarding my reconstruction of the accident disappeared. My name was mud around Allison and General Motors for some time after that. The NTSB, however, gave me an "atta boy" through Lockheed. This was the only known "whirl mode" experienced by a Herc and it resulted from a maintenance error. The irony of it was that a red grease pencil arrow had been drawn pointing to the missing bolt holes in the engine-mount-to-nacelle connection.

My first trip to Pakistan in April of 1972 was a bit unique. Due to their war with India, the U.S. State Department had embargoed them and it had not been lifted even with the war over. They had five Hercs which they had purchased earlier. I got called in one day and told they wanted me to go to Pakistan because they were having some problems with their Hercs. However, I must understand that Pakistan was embargoed. After some discussion, during which I was led to believe that the embargo was going to be lifted soon and Pakistan wanted to buy more Hercs, I said that I would go. I was told that no visa was required, but I found out differently.

When I filled out the entry card I had checked business rather than pleasure. A visa was required if you were there on business. Pan Am came to my rescue and scheduled me out on a flight the next day so I was granted an "over-night" entry. The next morning the Lockheed agent for Pakistan came and wanted me to go to Islamabad where the Hercs were based. I told him I was not going anywhere until I had a proper visa. He and I did a lot of running around on the back of three-wheeled rickshaw taxis and finally I got my passport stamped with a proper visa. Then I agreed to go on to Islamabad.

The first day when I returned from the Herc base and asked for my room key, it was handed to me with an envelope. In the envelope was a business card from Allen Shepard, Air Attaché in the U.S. embassy with a handwritten note on the back, "If I can be of any assistance." I really felt a lot better about being there — up to that time I had sort of felt like a criminal. I then realized that although technically not legitimate, my trip had tongue-in-cheek approval of the State Department.

It was a rather pleasant trip; I even went up to Peshawar, which is at the foot of the Khyber Pass on the Afghanistan border where three of the Hercs were based. With an escort I went up the Khyber Pass toward Afghanistan to a small village.

About half-way through the pass, a little old man squatted in a corner of a kind of lean-to shelter and drew pictures on Chinese silk squares using local herb colors to dye the fabric and white lead for the drawings, which he made using a sharpened stick or a nail. I bought twenty for about 30 cents each, except for one that he drew using white lead and then, while it was still wet, he would shake gold dust on it, let it dry, then shake off any loose dust. I brought them home and we gave them all away except the one I had watched him draw using a nail and the white lead paste and gold dust. We had it framed and it hangs in our house to this day.

All in all, it was an interesting trip but, boy, the debriefing I got when I got back to the plant. At the end of each trip into that area I could count on at least a day of debriefing when I returned. In the end, Pakistan bought ten more Hercs.

FIVE

Over the years I made several more trips into Pakistan and even up to Peshawar, but never again up the Khyber Pass as the civil war in Afghanistan heated up and Peshawar became an armed camp. Everywhere you went there were fierce-looking fighters out of Afghanistan carrying guns and ammunition belts resting and getting high on opium. In fact there was a very good business in manufacturing and importing guns that ran the economy in Peshawar.

Sudan had bought some Hercs and shortly thereafter had problems, so I went to Khartoum, Sudan. I thought that this was an interesting place located at the intersection of the Blue and White Nile Rivers. In those days I was walking four to five miles each morning and did so my first day there. After all, I could walk along the Blue Nile or the White Nile and remember a lot of the history that had occurred along them.

When I got back to my hotel I had the urge to urinate and proceeded to do so, but it was all blood. Talking about being scared; the next flight out was two days away. As I felt okay there was no way that I was going to a local hospital. After the shock, I sat down and reasoned that although there had been a breeze blowing, the temperature was over 110 degrees Fahrenheit. However, I had not really felt hot and had not really sweated much. So I decided to drink everything that didn't drink me first and everything cleared up overnight. Although I did my walk the next morning, nothing out of the ordinary occurred.

From then on I did take a bottle of water with me when I walked and upon returning home had a discussion with Louise. She had made an appointment with our doctor. When I told him my story he said that it was not at all unusual for athletes to do this if they didn't drink enough fluids. I made several trips back to Sudan after that but, believe you me, I made sure that I drank a lot of water. In fact that taught me a lesson, and I always drank a lot of water when I was in hot, arid places after that experience.

I always enjoyed Khartoum. The shops were different — you never knew whether an item was real or an imitation. When I had the Safety Briefing Team there, even after me warning them, a couple bought an $8,500 Rolex watch for $300. The watch ran for about a month before it stopped working.

I had an interesting experience in Brazil. They had an accident and I went down to investigate. It was rather straightforward — the pilot had absolutely screwed up, and I wrote it up as that. The problem was that the pilot was the brother of the Brazilian Air Force's general. All sorts of pressure was being brought to bear, but I remembered my instructions, "If it is the airplane's fault we will take our lumps, but if it is not, don't take the easy way out and blame the plane." So I went to my hotel room to call the plant for instructions. About half-way through our conversation the phone went dead. I looked out my door and, sure enough, there were three people in the hallway. I decided that I really did not want to go out, so for three days I used room service until the phone worked again and there were not any posted guards in the hall.

The plant had understood my problem and gotten hold of the State Department - at least I think that's what happened. I did not revise my report. There was never an explanation given me as to what action the plant took, I was simply told to "forget it".

It was times like this that I would have liked to have said, "Take it and shove it, I quit." But I never did. I really had no problem getting out of Brazil. Nor did I have any problems on later trips to Brazil. Although I was initially apprehensive, later I fully relaxed and enjoyed my visits.

In 1978 the USAF was flying missions from Incirlik Air Base near Adana in Southern Turkey to supply our listening posts along the Black Sea using C-130s on detachment out of England. In July there was a crash of a C-130 returning to Incirlik from the re-supply mission. It was a very hard accident to investigate as there was very little of the plane left after it impacted a cliff at an estimated 500 mph.

Eventually, I determined that all of the left wing was not present and a ground search ensued. After a day and a half we found the missing part of the wing and as I was inspecting it as it lay, up jumped the devil. I spotted a hole in the lower surface of the wing going into the fuel tank. Further investigation revealed that it probably was a bullet hole as the projectile had grazed a skin riser and passed through a tank vent tube that was inside the tank and as the upper wing skin panel was never located, may have penetrated it also or just disappeared in the resultant explosion. To me this indicated positive evidence of a wing fuel tank explosion as a result of a bullet fired from the ground.

As far as I was concerned, the cause of the crash had been determined. When the President of the Air Force Investigation Board called in to headquarters in the States with his report, all was well. But about four hours afterwards he received a call from Military Airlift Command Headquarters and things started changing. However, I insisted that the hole in the lower surface be cut out of the wing and sent to the FBI lab in Washington. This was done only after I threatened to leave Turkey. The section of the wing was hand-carried to the FBI lab in Washington for their inspection and testing. The results confirmed that the hole and damage to the vent tube was the result of a bullet.

I could understand the reluctance of the Air Force to accept this because of State Department pressure. Turkey was a "marginal" ally and was having very serious problems with insurgents trying to gain support to overthrow the government. But facts were facts, and I knew that Lockheed could be sued over that crash being caused by "design defects." The wing part was secured in the Air Force "evidence vault" at Wright Patterson Air Force Base.

Sure enough a suit was eventually filed, and at some point the plaintiff's attorney wanted to see the "evidence." So he got a court order but it was stipulated that Lockheed and their attorneys would be present. I was elected to be the Lockheed representative and we all gathered at Wright Patterson Air Force Base. Two armed USAF MPs opened three locked doors into the vault and brought out the part. I identified it as the correct part as I had placed a personal identifying mark on the tank side of the part in Turkey. It was photographed by both the Lockheed and plaintiff's attorneys under the watchful eyes of the USAF MPs. Afterwards it was returned to the evidence vault and locked up with all attendees signing that this had been done.

As the trial date approached, a court order was issued to the USAF to produce the evidence to the court. Would you believe the USAF issued an affidavit that the part had been misplaced and could not be located? As a result, after a lot of legal maneuvering, a token settlement was reached and the case was dismissed. One must never underestimate the power of International politics. Dismissing the case kept Turkey from being "embarrassed" in the Federal Courts of the U.S.

On April 4, 1982 I'll be darned if another C-130 on a re-supply mission didn't crash in the mountains of Eastern Turkey. I was in Panama at the time with the Briefing Team and received a phone call to "get to Turkey ASAP." I left the team and immediately started trying to get from Panama City to Frankfurt, Germany where a C-141 on a daily supply run was to take me to Incirlik, Turkey because the only internal Turkish airline was on strike.

On approach to Incirlik Air Force Base the pilot told me that they were holding a C-130 on the ramp with engines running waiting for my arrival. We landed, I off-loaded, bag in hand, and a sergeant grabbed me, asked for my passport and gave it to a person standing by him who stamped an entry visa, and immediately walked me to the C-130 with engines running. I got aboard and we took off.

Once airborne, I asked the flight crew where we were going. I was told Erzincan, where the investigating board had set up shop. They were a bit upset because they had been ordered to await my arrival and this would not give them time to make the flight to Erzincan and return to Incirlik before dark and they were not allowed to fly in this area after dark due to insurgent activity, which was not my fault.

We landed and I was greeted and taken to a hotel that had been setup as the Investigation Board Headquarters. There I found the Lockheed investigator living in Air Force contracted rooms. I immediately got a non-Air Force-contracted room and demanded that the investigator do the same — Lockheed pays their own way. The rooms were four dollars per night, not a big deal. Anyway, after a bit of discussion, I got the Lockheed employee settled down and demanded a briefing. I soon discovered that, although he was a retired U.S. Navy safety officer, he did not know his way around an accident investigation; in fact, I could not get much information out of him. I did discover that we traveled to the site via a locally contracted bus with a Turkey gendarme escort. I soon learn you did not wander outside the guarded perimeter without an armed escort.

The next day we went to the site. I had brought four rolls of 36 exposure film with me. The Lockheed Senior Safety Engineer told me that he had used all his film. He said that if I let him have the film, he would take pictures while I dictated my description notes into my voice recorder. I did this, much to my regret afterwards.

The crash site was a mess; there were 35 people on board. It was soon obvious to me that both wings had separated and the fuselage had fallen upright down the side of a mountain and crossed a small valley still intact until it hit the mountain on the other side at over 500 mph by my estimation and there disintegrated. However there were no signs of any wings, engines, or propellers.

This area still had snow, but it was melting rapidly and the local shepherds were walking their small herds from one small grassy area to another as the snow melted. I suggested that we enlist their support with small rewards for locating and notifying us or if necessary, taking us, to their "find." This was worked out and soon we were being led to plane parts all over the mountain. It soon became obvious to me that the right wing had been damaged by propeller strikes until it failed and then the sudden roll caused the left wing to separate.

The question remained, "How did the right wing get damaged to the point it failed?" Then one day in came a little shepherd boy to say that he had found some more parts. So off we went with him. There were a couple of pieces of a propeller blade and some twisted sheet metal. Looking at the sheet metal more closely, I identified it as the nose section of the right external fuel tank. Voilà, now all that remained to be solved was what had happened to cause the number three propeller to come off and strike the external fuel tank and then "walk across the wing chopping away the structure." Further inspection of the number three engine revealed that the front section of its reduction gear box was missing. The next day shepherds came in with the location of a "big part." An investigation revealed it was the number three propeller with most of the front half of the reduction gear box still attached.

By this time I was getting the hint of an idea. With everything loaded on Air Force flatbed trucks, we all headed back to Incirlik and the air base there. I wanted to dismantle the engine and reduction gear box oil filters but the decision was made to simply crate the engines and propellers and airlift them to Kelly Air Force Base. The only concession that I could get was that the teardown would await my arrival at Kelly.

When disassembly started at Kelly there were probably ten Allison people there and here I was all alone. My suspicions were confirmed. A planet bearing in the reduction gear box had failed and contaminated the oil filters and scavenge pumps causing them to fail. That engine would only run about one and a half minutes without oil circulation. With failed scavenge pumps the oil was not being returned to the oil cooler. Instead, it filled the reduction gear box and then overflowed. With no oil being circulated, engine failure occurred. Apparently the flight engineer was not monitoring his engine instruments as he should have been. Otherwise, he would have noted two things, the sudden depleting of oil in the tank and high oil temperatures in both the reduction gearbox and the power section.

When we had returned to Incirlik, our film was turned into the USAF lab for processing. The lab tech advised me the following day that there were no images on any of the rolls of film and showed me the clear film strips. I was absolutely flabbergasted and went to the Lockheed Safety Engineer who had gotten the film from me and had been taking the pictures as I directed. He then fessed up that he was not familiar with the camera he had brought and showed me what he had done. The lens cap had remained in place the entire time. I told him that he would have to explain to the plant why there were no pictures. His explanation was that the "Air Force confiscated them all." Right or wrong, I let it stand, but did tell his supervisor the truth. To my knowledge no action was ever taken, after all he was "a bright and shining star" in the Lockheed Safety Department which he eventually headed up.

Tom Roe was an attorney that was also an Air Force Reserve Colonel flying C-130s. He started suing Lockheed over C-130 accidents. This was greatly resented, as in his position with the Reserve he was privy to all Air Force communications on C-130 accidents. This was resented by both Lockheed and Warner Robbins Air Material Command who had engineering responsibility for the C-130. Finally someone high up at Warner Robbins got in touch with their U.S. Senator and a stink resulted. I was called in and had a statement taken. In the end, Tom was given the choice of suing C-130 accidents as an attorney or retaining his position in the Air Guard. He resigned from the guard with the statement that there was a lot more money in suing C-130 accidents than there was in flying in the Air Guard. I laughed at this as I knew his relationship with the C-130 Squadron where he had been the commanding officer for a number of years. People were nuts to think that after he left he would no longer have access to all the accident information he desired.

After a few encounters I woke up to the fact that Tom would always settle a case on very favorable terms. That was less costly to Lockheed than going to court, even if we were to win. I suggested to Lockheed that it might be in our best interests to encourage Tom to sue accidents before some "hot shot" attorney picked one up and really burned us. Although this was a practical suggestion, it was misunderstood by some and hurt my reputation. In some quarters I was accused of selling out to the attorneys. That was never "bought" by the powers that be and eventually such talk went away. It did hurt me though because of my loyalty to and love of the Herc. I never have and never will take any position that is adverse to the Herc. The rumbles about my disloyalty died away and time proved the wisdom of my suggestion.

The 1982 crash in Erzincan, Turkey became the case that I had been afraid of. Tom kind of dragged his feet on filing the lawsuit and a real "hot shot" attorney from Salt Lake City who flew his own private jet got most of the victims' families to sign on the line with an offer of "20% plus expenses for me and the rest to you and I will do all the work, you don't have to do a thing." Now talk about smart, this attorney filed the suits in the Texas superior court. . Both Lockheed and General Motors (parent of Allison) had plants in Texas, therefore it was all legal. The big problem was that the Texas superior court had produced, in some people's opinion, the greatest miscarriage of justice during the Pennzoil verses Texaco lawsuit a few years back, and the judge who had made the ruling was still sitting.

The lawsuit filed against Lockheed and General Motors alleged the usual "design deficiencies" and requested 100 million dollars as compensation for "loss" (of loved ones), and unspecified amounts for pain and suffering, etc. Lockheed's and General Motors' insurance companies almost panicked. I am not privy to exactly what went on, but Lockheed took the lead to try and work out a settlement — everyone was afraid of that Texas court.

Finally, Lockheed worked out a settlement of 40 million dollars to be split between Lockheed and General Motors. For reasons I am not privy to, Lockheed's insurance company paid with an understanding that General Motors insurance company would contribute half. Then up jumped the devil, for reasons unknown to me, General Motors' "Brass" decided to not honor the agreement. You can imagine what occurred next. Lockheed's insurance company sued General Motors' insurance company.

Part of Lockheed's insurance policy required Lockheed to provide "technical assistance as required." I had retired in 1984 but Lockheed's insurance company immediately contacted me setting me up as a consultant for them. The attorney representing General Motors was a first class jerk and dragged the case out forever by coming up with outlandish scenarios as to what caused the crash. It even went so far as them going to San Antonio, getting the engines storage cans opened "for study" and them actually mixing the number three and number four parts so that when they were re-canned they no longer contained the same parts.

However, I had made identifying marks known only to me on certain crucial items of the number three engine so that when the Lockheed's Legal team visited and had the engine cans opened for a short time I was very confused. But finally I got it all straightened out by comparing my marks with my photos. A big stink resulted with threats of suits for "tampering with evidence" and Allison finally admitted that in putting the engine parts back in their respective containers they "might" have gotten some parts mixed up.

I kept telling the Lockheed insurance people the key was the nose section of the external fuel tank and the identified propeller blade part from the number three propeller. A meeting was held with all sides and it was agreed that I would go to Incirlik Air Force Base in Adana, Turkey, find the "junk pile," and see if I could come up with the part of the external tank that I kept saying was so critical.

So in August of 1998 I went to Adana and Incirlik Air Base. USAF cooperation had been secured so I was met at the front gate by a major who turned me over to a master sergeant, giving him instructions to accompany me everywhere but not to assist me in any way. Well, we got to the salvage yard where the remains were piled. I photographed everything and then started looking. To my utter amazement, laying a couple of feet away from the large pile, I spotted what I had been sent to recover. I was flabbergasted and said so. The sergeant said, "Oh, after we had moved this pile of stuff down here, we were cleaning out the Accident Investigation Board room and found these pieces. I just brought them down here and threw them over the fence". I could have kissed him. I had anticipated a week or better searching through that pile. Anyway I photographed them as they lay, got him to pick them up and put them in the Jeep. We then went to Shipping and they were crated and sent in bond to Sacramento Air Base as agreed. To this point I had not touched the parts.

The shipment finally arrived at Sacramento Air Base and "the gang" assembled and opened the shipping box. I identified the parts, noting the propeller blade serial number, and then oriented the nose section of the external tank showing that it had been sliced off by a downward stroke, which isolated the cut to the number three propeller.

At the end of October, another meeting was held at Lockheed attended by some new faces from Allison, the Chief Engineer for the Herc engines, along with the head of Allison's Safety Department. After the meeting, at which nothing new occurred, the Allison people asked if it would be possible to see a C-130 airplane. After my shock, I said yes (remember I had a badge allowing me total access to the plant and escort privileges). So I escorted both sides out to the ramp where a new C-130 identical to the accident plane stood. I pointed out what I said occurred. Then came my second shock — I overheard Allison's Chief Engineer for the Herc engines say to the Allison contingent, including the insurance and legal people, "S—t, Ross is right. Why are you guys fighting?" I walked a bit further away and soon they said, "Thank you Ross , we've seen enough" and I escorted them out to the parking lot. I then went into the plant, straight to the Legal department and told the head attorney what I had overheard. About three weeks later I received a phone call telling me that "all has been settled, they folded." Again I received only a thank you, but I had made a lot of money from this case after retirement. From occurrence to finalizing, this accident had taken sixteen years and eight months to resolve.

One Friday in September of 1972 at about 3:00 P.M. the phone rang on my desk and I answered. A voice said, "Cut a TARE (Lockheed's travel authorization form) for Saudi Arabia." I said, "Who in the he—is this?" The voice explained that it was the Chief Engineer and I had better get the lead out and get over to the Vice President of Marketing and get the details. I went on the double.

It seemed that Lockheed had just had a phone call from a Marketing representative who had been in Saudi Arabia and had flown out to Beirut, Lebanon to be able to call the plant and request my technical assistance in attempting to sell the Saudis several aerial refueling versions of the C-130. The representative would be at the Phoenician hotel in Beirut until Sunday evening when he had to fly back to Riyadh. Talk about paperwork flying; I really didn't know that Lockheed could do it. The Traffic Department was getting me a flight schedule, the Finance Department was getting together $9,000 and I was gathering information and data on the U.S. Marine version of the KC-130. I called my pastor to get a statement that I was a Protestant and in good standings with his church, the county Sheriff for a statement that I was not wanted or suspected of any crime, arranged to pick these statements up Saturday morning, and got Louise to dig up a certified copy of our marriage license. It all came together and Saturday night I was on a flight to Beirut via London.

The flight from London to Beirut was late and by the time I landed, cleared customs, and taxied to the Phoenician hotel, the Lockheed Marketing representative was standing by the door, bag in hand. My briefing went something like this, "In the morning a man from Cedars Travel Agency will call you and come get your passport. He will get a Saudi Arabia visa and bring it back to you — pay him whatever he asks as tomorrow is a holiday and he has to "backdoor" your visa. Then the next day fly to Riyadh and I will meet you." Boy what a briefing; I knew just exactly what was expected of me in Saudi Arabia. To make a long story short, things worked out just as he said. I had to pay the Cedars Travel Agency representative $300 but did get a receipt. The next day I flew into Riyadh and as I came through Immigration, I had to surrender my passport and was issued a Work Permit. Although I didn't like this, I had no choice. I was met by the Lockheed Marketing representative and his assistant. They took me to the hotel, which had been the home of the old Saudi King Faisal's harem before he outlawed harems late in his reign. There were about 75 rooms in this building. That evening I finally got a chance to talk with the Lockheed Marketing representative and get some idea as to why I was there.

So the next day we climbed aboard a Saudi Arabia C-130 (with a mixed Lockheed and Saudi crew) and off we went. At 25,000 feet two Saudi Lightning fighters were to join us and I was to get them positioned in the approximate position they would be in if the plane had the Lockheed adapted "Hose and Reel" refueling pods attached. The Saudi Arabians wanted to test if the fighter planes could position themselves near the Herc for refueling without stalling out. I instructed the flight crew as to what power settings to use and sure enough, at 25,000 feet two British-built Lightning fighters appeared. I got them positioned for refueling via a radio link and this maneuver was repeated several times. Only once did a fighter stall and have to break off and its pilot radioed that this was his fault. After this, I thought that my job in Saudi Arabia was complete and I could go home. But no, I was told by the Marketing representative that a Royal Prince (they didn't tell me he was the equivalent of our Secretary of Defense) wanted us to go on a picnic in the desert and they had accepted. I had no choice; after all, I didn't have my passport. So I sat and "twiddled my thumbs" being ready to go at a moment's notice.

We all boarded some big Mercedes buses with big tires and went out into the desert to an oasis where several Mercedes trucks were parked with a tent all set up. After a bit the "picnic" was all spread out and I was appointed to reach into the common pot and get a bite of meat as the honored guest. I did and came out with the sheep's eyeball — immediately everyone started clapping and cheering — and I had no choice but to put it in my mouth amid more clapping and cheering. I made chewing motions and swallowed the thing whole. Afterwards I was suspicious that I had been set up.

The following day my passport was returned with notice of a reservation on a flight back to Beirut. From there I got hold of Cedars Travel Agency and had them schedule me back home. The sheep's eye episode preceded me to the plant as the Marketing representative had flown to Beirut with me to "make some phone calls." At some point afterwards I found out that our host on the "picnic" really was just sizing me up. I guess I passed inspection since Saudi Arabia wound up buying about 75 Hercs by the time I retired.

In September of 1980 an unqualified Saudi crew had flown into Medina to pick up the King's Guard. The flight from Jeddah to Medina went well and the King's Guard was loaded. On take-off the plane got airborne and at an altitude of about 50 feet, nosed down and impacted the ground and with the resultant fire, all aboard perished. I was able to investigate the crash since the airport in Medina is outside the walls of the Holy City.

One of the first things I learned was that the Royal Family had already paid the families of all who had perished a "handsome sum." Investigation of the crash site revealed that all engines were "burning and turning." The propeller blade angles were in the take-off position upon impact, confirming that all engines were operating normally. A check of the flap jack screws revealed that the flaps were in the fully retracted position rather than the take-off position — case solved, or so I thought. The President of Lockheed Saudi Arabia wanted to see me.

It seemed that the Saudi Air Force wanted a cover up and had someone shoot a bullet into one of the engine turbine cases as "proof" of "rebel activity". Well, the engine turbines rotated at 13,800 rpm and although there was a bullet hole in a turbine case, with the resultant metal distortion, no turbine blade was deformed, which was proof that hole was made after the turbine had stopped rotating. I stood pat against the President of Lockheed Saudi Arabia's ranting and raving about the money Lockheed made in Saudi Arabia, etc. Things got pretty hot and I finally called my chief engineer; he confirmed that I was to "call it as you see it." My final parting word from the President of Lockheed Saudi Arabia was, "I'll have your ass," with me replying, "You don't have what it takes to get my ass." I went to the airport and flew home.

The debriefing to the executive committee that followed was not one of my more pleasant experiences. Although there were a lot of snickers around the table as I described my parting, I was told that I could have been more "diplomatic." A few years later this guy became President of Lockheed Georgia and I was told one of the first things he did was state that he wanted to see Ross Holdeman. Whether this is true or not, I don't know, someone may have just been poking fun at me. At any rate, I never ran into him, nor did I go looking for him while he was President of Lockheed Georgia.

At the end of April in 1982 Bolivia encountered a crash and Lockheed decided to send a Safety Engineer with me. I had been making noises about retiring. Unfortunately, it was the same safety engineer I had encountered in Turkey. I let it be known right up front that this was my investigation and he was welcome to tag along. It was a rather straight-forward accident. The pilots were trying to "sneak" into the Bogotá airport under the cloud cover, couldn't make it, abandoned their approach, climbed out for another try, and missed the top of the surrounding mountains by about 50 feet at the 18,000 feet level.

The first day it was a very hard climb to the impact site. At about the 14,000 feet level I stopped to rest and get my breath and a native woman about six to seven months pregnant passed me like a freight train would a tramp. I did a bit of investigating and found that a four-wheel drive could get to the top in a round-about way and we could walk down to the impact site. So I rented a four wheel drive Jeep and guide. We got to the top and I sent him back with the Jeep to the bottom of the mountain where the trail started. All went well until weather moved in and it started snowing; with the wind blowing the temperature plummeted. I said, "Let's get out of here."

The young Safety Engineer wasted no time and took off down the mountain on the double. There was no way that I could keep up and had a very hard time getting down the mountain to where the Jeep was waiting. I gave the Safety Engineer what for and when we returned to the plant I informed management that I would not travel with him again. I was never asked to do so, but a Vice President did ask me over a drink what had happened and I told him. I thought that the "powers that be should know" as he was a "rising star." Later he was passed over for promotion. At some point after I had retired I ran into him in the plant and he said he wanted to talk. He asked me if I had anything to do with him being passed over for advancement and I answered that I had never been asked about a recommendation. Actually, about five years later he was promoted.

The U.S. Navy had a requirement for better Carrier Onboard Delivery (COD) aircraft to bring aboard mail, fresh fruit, restocks of ammunition, etc. with the carrier at sea. The sort of "piggyback" system they had been using was no longer doing the job. The decision was made to try a C-130 for this job. This was totally a Navy project with Lockheed supplying requested information and sort of cheering the effort on. The Navy decided to use the USS Forestall and a USMC KC-130F. The only modifications to the plane were removal of the external tanks and installing the Anti-Skid system that was used in current production of the C-130 aircraft (it had some improvements over the system that had been delivered with the KC-130F).

The tests started in October of 1963 and were being accomplished with a Navy crew supplemented by a Lockheed Flight Test Pilot as advisor. Over a period of several weeks a total of 29 touch and go landings, 21 full stop landings and take-offs of an airplanes with gross weights of 85,000 to 135,000 pounds were performed. I was on the Herc for one landing and one take-off.

Part of the landing procedure was to reverse the propellers while about two feet off the deck and they had experienced some propeller rpm surges and wanted me to take a look. All this was hush hush — why I never understood. In about three months the test airplane came to Lockheed for its annual inspection and painted on the nose of the plane was a large sign that read, LOOK MA NO HOOK. I happened to see the plane land at Lockheed and called Security. Everyone got in a dither and they came to the ramp and covered the sign with masking tape. In a few days this was removed and during the annual inspection, the airplane was cleaned for re-painting and the sign went away.

Although the tests proved that the Herc could make the COD mission, they also showed that if the plane experienced a problem while on deck and was unable to depart, no other plane could land or take-off from the carrier as the Herc "just took up too much room." Although Lockheed made an effort of studying wing folding for the Herc, the Navy contracted with Gruman to modify the fuselage of a current in production submarine hunter plane and turn it into a COD.

In August of 1976 the Venezuela Air Force was to take the 56 members of the University of Caracas choir for a six week singing tour of Europe in one of their Hercs. The first stop was to be to an overnight stop at Lodges Field on the Island of Terceira in the Azores. Lajes Field is a large airport built by the British and U.S. during WWII as a layover site for flights from the States to Europe. Therefore there were many Officers Quarters buildings there. These are now operated by the Portuguese with rooms running around $5.00 per night and this makes a perfect night layover facility. The Venezuela Air Force had made arrangements to land there and billet the choir in these facilities.

Upon arrival the weather was "socked in" with the airport closed. After circling for around an hour the pilot decided to try and land using a Ground Controlled Approach (GCA) approach. This is a technique in which a person in the tower looks at two radar screens, one for altitude and one for the plane's ground position compared to the runway, to talk the pilot through the landing. Two attempts were "waved off" (aborted) by the GCA Radar operator who was a well qualified USAF Sergeant. On the third attempt the GCA operator again called, "abort!" but the pilot continued to descend hoping to see the ground. This information was from the GCA and Tower voice tape of which I had secured a copy.

Obviously the pilot never got a visual and applied power to go around again but did not get the aircraft's sink rate arrested and impacted the ground at about 140 miles per hour. He was not even aligned with the runway but off to the side by about half a mile. Lajes Island grows wine grapes in small patches with volcanic rock walls around each patch. The volcanic rocks used to build these walls were picked out of the fields to allow for cultivation. The airplane touched down and started plowing through these walls, disintegrating until there was literally nothing left. Those volcanic rocks really did a job on that plane. The pilot had made a terrible mistake but in some ways I had to sympathize with him since nowhere else in his fuel range could he find facilities to overnight 56 people, plus six crew members.

Venezuela called requesting assistance. You can imagine their shock over loosing 56 university students as well as the six man flight crew and plane. It was a national tragedy. I managed to get to Terceira commercially by going to Ottawa, Canada, taking a Portuguese National Airlines (TAP) flight to one of the other islands, then a "puddle jumper" over to Lajes Air Base.

Upon arrival I was met by several Venezuelans and a USAF captain. I introduced myself and told them that I was there to assist in the accident investigation. They said essentially "you investigate and report." I asked the captain about transport and lodging and was advised that he would take me to the billeting office, see that I got a room, take me to the village of Lajes, and direct me to the impact site where we would agree on a pick-up time. The USAF was providing assistance that included bringing in a C-141 to fly the remain bags back to Caracas. The captain said that he would provide transport from billeting to the village fountain and back to billeting but could not be involved in my investigation. A cafeteria was available as was the Officers Club complete with dining facilities all within walking distance of my billet. After getting a room and changing into "work clothes" he took me to the village fountain and showed me the way to the impact site saying he would return at such and such a time.

The impact site was total destruction. All human remains had been removed and bagged awaiting a USAF C-141. The plane obviously impacted the ground with a high sink rate and forward velocity. It tore through those rock walls and was chewed up as it slid. I started doing my thing, gathering physical evidence such as flap position, engine power at impact, landing gear position, etc. After a couple of days of work, I started back to the village fountain to await pick up. I was tired and had started there about three to four hours early, even thinking I might walk back to billeting.

Since arrival I had noticed some two-wheel carts with very large barrels with a long wooden tongue for a horse to pull attached. This afternoon, as I came into the square, I noticed one of these "wheeled barrels" beside a wall and went over to investigate. While looking into it, a door in the wall opened and out came a little fat man. A lot of arm waving ensued and he took me by the arm back through the door in the wall. I found myself in a courtyard with all sorts of things that obviously were used to process the grapes. A large circular vat about four feet high stood in the middle and it became obvious that was where the grapes were stomped. I pointed to him and stomped my feet, he said, "No, no, Minha esposa Sonora fazer isso," meaning that his wife

Sonora does that. He offered me a drink from one of the wheeled barrels and he drank also. After a couple of large cups, I simply backed up to the wall and slid down into a sitting position and slept.

When I awoke it was past time for my pickup. I had a pretty good "buzz" but started walking. Eventually a U.S. military vehicle came along and I flagged it down and got them to take me to my quarters. That wine was mighty tasty but it sure was potent. During the rest of my stay there, I assure you, I never inspected a two-wheel barrel again.

I went to the tower and got a copy of the radar and voice tapes and actually did a nice job of plotting the three approaches, both vertical descent and horizontal positions, against the runway threshold. The C-141 came in and took the body bags away and after about five days a meeting was held with the USAF, the Portuguese Base Officer, Venezuela Air Force, and their equivalent of our State Department. I presented my factual report without any finger pointing. Afterwards I had an interesting question posed by the Venezuela Air Force representative. He asked me to theorize as to the cockpit thinking. I told him that it was my personal opinion that the pilot, after lingering for over an hour waiting on the weather could not reach Lisbon, and Lajes was the only place within his fuel range that could billet his crew plus 53 choir members. He had no option but to try and get in, but his approaches were badly executed, maybe from gusting winds, and he had simply descended too low trying to see terrain. He had given up on that approach and applied full power to go around and try again, but the sink rate of the airplane could not be arrested before ground impact. I never used the words "pilot error," and in the end I was thanked and started making arrangements to get back to Lockheed. I never heard another word about the accident or report, I don't even know if Lockheed received a word of thanks or not, however the Venezuela Air Force did buy several more Hercs and I visited the Herc Squadron in Venezuela on a few occasions.

Six

In the early 1980s a big controversy broke out over whether the "stretched" C-130 would enter "fin stall" or not. Fin stall required slowing the airplane and holding the wings level while applying left rudder. As the nose of the airplane swung out from the center of gravity to about 25 degrees left in yaw, the wing blanked the airflow to the vertical fin and rudder rendering them ineffective. The airplane would then roll left, pitch down, and as soon as air speed increased, recover to normal flight. Since this was an abnormal flight situation, it was not considered to be a defect in the airplane. However, the controversy continued over whether or not the 120 inch fuselage length increase in the "stretched" C-130 could cause unrecoverable fin stall. The United Kingdom had purchased some "stretched" C-130s, so a Lockheed Vice President asked if Flight Operations would take a "stretched" Herc and settle the controversy. The problem was that the test was to be done without installing pitch, roll, and yaw, test instrumentation due to costs.

    The chief pilot agreed (dumb decision) and off they went with the chief pilot in the left seat and the best, most savvy pilot at Lockheed (in my opinion) in the right seat as copilot. After a couple of tries with no fin stall they yawed really hard to the left and suddenly the airplane flipped on it's back and started an inverted spin. The chief pilot panicked but the copilot stayed calm and finally got the plane back under control and they brought it home with a shaken crew.

Because this was a commercial version of the Hercules, it had a flight recorder but a flight operations cover up was in progress; I demanded the tape from the flight recorder. It finally took a direct order from the Lockheed president for me to get the tape. I was so insistent because I remembered the Arkansas crash where the plane had spun and damaged the internal structure of the wing. When I got the tape, I found that flight's portion had been crumpled up but I was able to straighten it out and what I saw shook me up.

As I read the tracings on the tape, I learned that the plane had inverted, spun several turns, and during recovery had generated 8.2gs even though it is only rated to withstand a maximum of 4gs. I demanded that a complete structural inspection of the plane be made, including opening the fuel tanks. When they were opened all hell broke loose. Just as I had suspected, all the internal structure had been torn loose from fuel being violently forced toward the wing tip. This discovery led to many other inspections including having to have the engines and propellers overhauled.

I said something to the copilot about pulling 8.2gs and his response was, "Hell, I was counting the leaves on the trees by that time." I remain amazed to this day that the g-force of the pull out didn't tear the wings off and crash the plane. After many hours of repair, engine and propeller overhauls, and further inspections, the plane was declared airworthy and sold to a foreign operator. About six months afterwards, the local service representative called the plant about wrinkles and "oil cans" in the vertical fin. Further inspection revealed that the plane had also suffered internal structural damage. Lockheed fixed it and to my knowledge the plane flies to this date. The copilot on that plane was small of stature but one of the best pilots I ever had the pleasure of knowing, eventually he became Lockheed's Chief Pilot and retired from that position.

At one point we had a crash in Lagos, Nigeria and my assistance was requested. Upon arrival I discovered that the impact site was void of any wreckage. Local inquiry revealed that the local witch doctor had it all in his yard. It seemed that he had paid the locals to remove the remains from the impact site and bring it to his yard before the officials arrived on site. So when I arrived, I was expected to determine why the plane had crashed, but all the remains were on the local witch doctor's property in a big heap. He and the local military were not on the best of terms, so they were forbidden access. After a very frustrating day I thought about it and the next morning I had my driver take me to the witch doctor. After about an hour of negotiating, it was agreed that I could examine the pile for $20 U.S. down and $20 U.S. per day, and he would provide people to move what I wanted moved. This worked out well and after two days of investigation I was ready to write my report. I did and returned home. You should have seen the commotion when my expense account hit Accounting with a charge of $60 paid to a witch doctor and no receipt. It took about a month before that got settled. I threatened to recall the expense statement, redo it, and include the $60 in other line items where they would never see it. The C-130 program Vice President finally settled the argument by telling Accounting to "approve as submitted."

On July 1, 1976 I received a call from an Israeli C-130 flight engineer asking for a reminder of what I had told him about a way to get more power out of the engines in an emergency. The power output was controlled electronically to a maximum of 4,200 horse power, but by rapidly moving the throttles from "idle" to "take-off," the electronics would be in the process of increasing fuel to the engine, and if you moved a switch to lock the electronic control in that position before it reacted, you could produce more than the 4,200 horse power that the electronics allowed. So in response to his question I retold him how this could be accomplished but reminded him it certainly was not an approved maneuver. A

On Saturday, July 4, 1976, I was mowing the yard when Louise called me in for a phone call. It was an Israeli flight engineer who reminded me of that one time he had a Herc engine that was "hot starting" and I had hit the fuel control and gotten a normal start. A hot start occurs when an engine is being started and the fuel is ignited before the engine is turning over fast enough. He wanted to know where and how hard to hit the fuel control. He said that they were away from home on a maneuver and really needed to get an engine started immediately. I told him but reminded him that it was no "guaranteed cure."

The next day I read of the Entebbe raid by the Israeli military, where they had taken three Hercs and flown to Entebbe, Uganda in Africa, landed at night, found the hijacked prisoners, rescued all of them, and flew them back to Israel. I finally got the story. That phone call for urgent help had come from the Entebbe runway. All the questions that I had been asked at the end of June and first of July relating to increasing engine power output, etc. all related to preparations for this raid. The malfunctioning fuel control that I was called about was on the plane that had all the rescued people already on board and the engines were being started to fly back to Israel.

The Israeli raiding party had a high-flying converted Boeing 707 full of communication equipment covering the operation. The people on the ground had radioed it and had it patch them through so that the flight engineer could talk to me. All's well that ends well and all three Hercs returned to Israel with all the hijacked passengers. The only thing on their list that they didn't accomplish was stealing Idi Amin's personal civilian version of the C-130. They just could not find it at the airport. When I learned of that intent, it explained all the questions I had fielded about the differences between the civil and military versions of the Herc. I did report the July 4th contact from the Israeli flight engineer to the "powers that be" at Lockheed but no questions were asked.

In November 1978 there was a large thunderstorm in the Charleston, South Carolina area and a USAF Reserve C-130 was approaching for a landing. They ran into the thunderstorm with extreme turbulence and wanted to climb above it but were not allowed to by Air Traffic Control as an Eastern Air Lines 727 was also approaching at a higher altitude. Shortly after this request was denied, the crew radioed that they had "just taken a lightning strike on the right wing and were going to turn right to evaluate." After that transmission radio contact was lost with the plane.

The plane was found in a swamp just to the right of the runway's approach path. Investigation revealed that it had impacted nose down and was sinking. In fact, it was sinking faster than we could dig. Anyway, I measured the width of the impact site and found it short of the Herc's wing span. This led me to believe that the cause of the accident would not be found at the impact site but out there somewhere in the swamp. I felt sure that a large piece of wing was out there somewhere, but the USAF Board of Inquiry President insisted we dig in the hole. Now you talk about a muddy mess every evening, that was me. I tried to protect the rental car I had but didn't have much luck.

After about three days, I convinced the Board President to go to Jacksonville and get a print out of the radar contacts with the plane from Air Traffic Control. He did, and upon returning, I was reading the copy and noted that just before the image disappeared from radar contact, there was suddenly two image returns in essentially the same location. This convinced the Board to start a ground search back along the flight path.

We had about fifteen people spread out staying within easy sight of each other. After about four hours of sloshing through the swamp, about 25 feet of the right outboard wing was located. The location was flagged by climbing a tree and tying large red and yellow streamers to its top. The next day a chopper was secured and back we went to the wing. I supervised a sling being placed around the wing portion. The chopper picked the part out of the swamp and placed it on hard ground. The evidence was irrefutable — a lighting strike had hit the right wing with enough energy to burn a hole in the upper surface of the wing and ignite the fuel which exploded, destroying the internal structure of the wing. When the pilot decided to "turn right to evaluate," the weakened wing could not support the increased load and failed.

An interesting sidelight to this accident is that when the Board President left the site to go to Jacksonville to get the radar plot he left me in charge — why I don't know, since there were other military members of the board present. But to make a long story short, about the middle of the morning one of the perimeter MPs radioed me that a news man wanted to come in to the secure site. I didn't really know what to do, so to be safe, I said no. Later that afternoon, I noted a light plane circling the site at a rather low altitude but ignored it. That evening in my motel room as I was watching the evening news, there were aerial photos of a plane crash, and the description was so inaccurate that I thought another crash had occurred and I called the plant to check.

That put the wraps on this accident and I got reservations back to Atlanta, checked out of the motel, and drove by the Investigation Board Room to say my good-byes. Upon entering I noted that something was wrong. I was then told that another Herc had crashed at Ft. Campbell, KY and the pilot was a very special friend of this board's President. I called the plant but was told to come back home and we would make plans on Monday. I did so since this was Saturday and I wanted to go home and at least get laundry done.

Monday I went into the plant and by noon I was on my way to the Atlanta Airport bound for Ft. Campbell. This was an accident that never should have happened. The plane was on a training flight and was coming in for a landing. On final the pilot called the tower and said that they had "something we want to check out so are breaking off the landing and leaving the pattern." Shortly it nosed down and impacted the ground. Here was a very experienced C-130 pilot with a relatively inexperienced copilot. Although the impact was fatal, the aircraft was traveling at about 400 feet altitude and at a relatively slow speed. For some reason there was no fire so we had a "clean" impact.

The investigation proceeded methodically, determining that all flight controls were functional, that all engines were producing power, and that all the propellers were attached. As they approached the landing area, the pilot had radioed the tower that something was not right and they wanted to check it. Finally the clue — one of the propeller's oil filler cap was open allowing the propeller oil to spew out. Further checking revealed that normally this cap was closed and locked by inserting a locking pin through a hole in the housing and through the cap. It was a pin that had retractable balls in its end so that, without depressing a release button, the balls would not retract, thus preventing the pin from "backing out" of its locked position.

Some time in the past it had been discovered that some mechanics, in servicing the propellers, were not firmly seating the locking pin and the locking balls never extended. To counter this, Hamilton Standard had issued a caution to propeller mechanics and instructions to counter bore the locking hole at overhaul so that the locking pin did not have to be so firmly seated for the locking balls to extend. Although this propeller had been overhauled twice, those instructions were never carried out. As this airplane was approaching landing, that locking pin came out, the cap opened, and propeller oil spewed out and control of that propeller was lost with the "Propeller Low Oil" light illuminating. Without oil the propeller could not be feathered. They were at a slow speed and with drag increased by a wind milling propeller, the airplane stalled and crashed. The most minor things can, under the right circumstances, cause the loss of a plane and lives of the crew.

I had always been fascinated by the fancy Greek ouzo liquor bottles shaped like old temples. In May of 1983 I was in Athens and was scheduled to go home from there, so I went out and bought eight different bottles. I didn't care about the ouzo. I wanted the bottles and I didn't know whether I would be back in Greece before retiring or not. As I was about to leave Athens, I got a call from the plant to go to Saudi Arabia for some big problem with one of their Hercs. It hit me that Saudi Arabia did not allow alcohol to be brought into the country so I was about to lose my nice bottles when an idea hit me. I called the hotel concierge, explained my problem to him, and told him that if he would get some empty bottles, we could pour the contents of my bottles into his bottles and he could keep the liquor. I would wash the bottles very well and keep the originals. He jumped on that deal like a rooster jumping on a grasshopper. After emptying the contents of my bottles into his empty bottles, I washed mine out leaving no liquid or odor.

I landed in Riyadh Saudi Arabia and proceeded to Immigration and Customs. When he opened my suitcase, the inspector gave me a hard look from under raised eyebrows as he reached for a bottle of what he was certain was contraband. I just smiled as he inspected, twisted, turned and sniffed every bottle before losing interest. I did get every one of those bottles home and we still have them today. I was always pleased with that accomplishment as well as the bottles and that was my last trip to Greece for Lockheed.

In April of 1978 I had the Safety Briefing Team on a tour of Herc operators in Canada and Alaska. We were between flights in Seattle where I bought a newspaper and saw that a C-130 had crashed at Sparrevohn Air Force Base in Alaska. I called the plant and discovered that they had been trying to locate me and I was to go to Anchorage immediately. Well, I could not get there any faster than my current schedule, but when I arrived, I made contact with the USAF at Elmendorf Air Force Base, left the Briefing Team and was on my way to Sparrevohn AFB which was one of the Distant Early Warning (DEW) Line radar bases that covered the Northern globe from Western to Eastern Russia. I had been there once before on a re-supply mission with Alaska Air Lines.

The proper approach to this base was to line up with the runway, full flaps, flight idle power, and literally "fall down the side of a mountain" to the end of the runway on the other side of the valley, touch down, and apply power to climb the hill to the end of the runway, which ended at the base of a 1200 foot vertical bluff. I had been there and done that and was not happy about going back, but here I was on a Herc heading there. The approach information specified that after full flaps and flight idle power had been selected, if one lost site of the runway, one was to brake left, apply take-off power, move flaps to 50% and after regaining airspeed and visibility, climb out of the valley and try again. The approach information also warned that under no conditions should one continue his approach if he lost site of the runway. On my first trip in there with Alaska Airlines I was told, true or not, I don't know, that there were over twenty crashed airplanes around that runway.

The witnesses who saw the Sparrevohn plane crash I was to investigate said it started the approach down the mountain and then it hit a snow shower. The next thing they saw was the airplane turning right trying to out climb the mountain grade in a nose up position when it hit the mountain. They used a snow cat and eventually reached the impact site and removed the victims.

We were not allowed to go to the site until an armed perimeter was established as the bears were just coming out of hibernation with their cubs and were hungry. There was concern that the smell of blood and flesh at the impact site would draw predators.

Because my trip started out as a Safety Briefing Team trip I only had suits for clothes. I borrowed a flight suit and a pair of oversized boots from the station that I wore over my dress shoes onto the impact site. The first thing that I noticed was that the landing gear was down and the second thing was that the flaps were retracted. What a go-around configuration — all wrong. To have had any chance of out-climbing that mountain after he turned the wrong way, the pilot should have had the flaps at 50% and gear up. I don't know whether he could have made it even then. It would have been close. Now my job was to prove that the airplane systems were all functioning normally and the accident was the result of pilot error only.

At my retirement the Pentagon sent a Colonel down with a large picture of me swinging a sledge at the Sparrevohn impact site, trying to get a propeller dome open to confirm blade angle at impact and it was presented to me at my retirement party. The picture contained glowing statements by two Colonels who had at separate times been head of the C-130 Safety Department for the USAF, thanking me for what I had taught them, and how I had enhanced the USAF's knowledge of the C-130 and contributed to its safety record. I am very proud of that memento which hangs in my office today.

On February 1, 1975 an Air Force Reserve unit had a Herc crash at a WW II air base being used by a Navy reserve organization just outside New Orleans. I was sent to provide technical assistance to the Accident Investigation Board. This accident ranks high on my list of completely stupid, self-induced Herc accidents. However, it also resulted in my initiation to the world of "Ambulance Chaser Attorneys," the court system, and insurance company philosophies, that I would live with for the next 35 years.

This accident occurred because the flight crew did not follow the flight handbook. During flight they encountered an engine malfunction and correctly pulled a circuit breaker removing power from that engine component, which was required only during engine power acceleration from the "ground idle" to the flight range. It also was an item listed in the "no go" list for take-off. If that item was not functioning, the pilot should not attempt to take-off. The flight proceeded normally until they landed and decided to do a "Touch and Go." That engine component was not needed to stay in flight but it was definitely needed to help get the engine back up to speed after being in the "ground range." Because the power was removed when the breaker was pulled, it couldn't do its job and the number four engine did not power up. With the other three engines producing take-off power, the airplane was still below minimum engine out control speed and started going left due to asymmetric power. The flight engineer unbuckled his seat belt and stood up to pull the fire handle to shut down the number four engine and reduce its "wind milling drag." Before he could do that, the airplane impacted trees on the edge of the airport. All the other crew members were still buckled in and survived. This was so blatantly crew error that we at Lockheed were shocked when a "notice of suit" arrived.

The plaintiff's attorney, a Mr. Cooper from Baton Rouge, and I took an immediate dislike for each other. As his deposing me neared, I told the Lockheed attorney, "If I stand up and reach across the table toward that SOB, grab me." He said, "Don't do that. Remember that you can't get mad at a person unless you are looking at them." When deposition time rolled around, I sort of turned sideways and proceeded to answer his questions. He eventually maneuvered around to get in my eyesight, and I turned the other way. This occurred several times and he was getting more and more frustrated and finally said "Damn you, Ross Holdeman, look at me." I replied in a very calm voice, "Sir, the law says I have to answer your questions, but it doesn't say I have to look at you." He almost exploded on the spot. This was all being taken down by a court reporter, which made it all the more fun. My deposition didn't last very long after that. The Lockheed attorney was right, I never did get mad, but I sure frustrated that attorney.

About three months afterwards I went to New Orleans for the trial. The Jury selection was something else; it was a six person jury and only one had graduated from high school. Here we were, all prepared to argue asymmetric forces, reasons for flight handbook procedures, and the USAF requirement that they be followed. Anyway, after taking a look at the seated jury, eye balling the widow and her six small children seated on the front row who cried on cue, the Lockheed attorney requested a "bench conference." The Lockheed insurance people and all the attorneys gathered at the judge's bench. After about twenty minutes, the judge announced that Lockheed had made a settlement offer and it had been accepted and excused the jury.

I came unglued, condemning the court system. The judge informed me that "justice prevails" and I spouted off, "It sure as hell didn't in this courtroom today!" and with that one of the Lockheed attorneys grabbed me by the arm, dragged me out of the courtroom, out of the courthouse, and down the street. I got a reaming that was the father of all reamings, even with my fighting back. He finally got me to a bar and started feeding me bourbon. I finally cooled down, went to my hotel, and called Lockheed Legal. They had already been called and simply said it was okay.

When I got back to the plant I did a lot of ranting and raving as to why should we investigate accidents, and didn't Lockheed have any say as to whether a case should be settled or not. The end result was that I created such a resentment toward Engineering spending money to investigate accidents and supporting the defense of legal suits that the Legal Department got the insurance company to come to the plant and meet with all Engineering in one of the theaters to try and cool things off. They did and fully explained that any time they could settle for less money than the projected cost of a trial, they would settle. They were insuring Lockheed against liability loss, not defending a plane's design and construction reputation. This did cool things off a bit and I learned to "keep my cool," even after Cooper sent me a TWX that said, "I told you so, ha, ha."

I did get revenge later. We lost a plane in Arkansas and Cooper jumped on it, splitting it with Tom Roe. This was one of those cases that Lockheed and the insurance company agreed to defend to "take the wind out of some attorney's sails." I had investigated that accident also and Cooper was in a box. He knew I would do all I could to frustrate him, but I was the primary defense witness and he had to depose me, which he did, but rather gently, with me still not looking at him. Then we got to court and after my testimony, he started his cross examination. The Federal judge was a woman and we had a jury that appeared to be intelligent. During pre-trial maneuvering we had gotten the judge to rule that the Air Force's final report on the accident was inadmissible as they really are confidential and are generated in the interest of safety only. Therefore all information contained in it is considered absolute, with no information held back for fear of legal action. Remember that Tom Roe still had friends in his old C-130 unit and got a copy of the final report. During two or three hours of cross-examination Cooper had provided me numerous documents to identify and be questioned on. Finally, he handed me one and I started to read it and recognized it as a copy of the official final report. I turned to Cooper and said, "I won't read this because it is classified information. He called on the judge to order me to do so and with that I replied that it was my understanding that she had ruled that the final report was inadmissible, and this was a copy of the official final report. That really woke the judge up and she ordered everyone into her chambers. Cooper got a reaming like I had never seen before and a $2,500 fine. In his closing argument he referred to my testimony saying that it "should be discounted because Mr. Holdeman loves that plane and everyone knows love is blind." That almost brought the house down. The case finally went to the jury and within an hour they returned with a not guilty verdict for Lockheed on all counts. Revenge was sweet, and this marked the end of "Mr. Cooper from Baton Rouge" suing C-130 crashes, but unfortunately, the plant would not let me send Cooper a TWX.

Although I was retired, a Southern Airways plane on a contract logistic flight for the USAF crashed on take-off from Kelly Air Force Base in San Antonio, Texas. Lockheed's insurance company told me to stay on top of this and so advised Lockheed. Although there were only three deaths, the plane went into a hanger that contained twelve F-15 fighters from a South American country that were there for overhaul and exploded, destroying all the fighters. As I recall, the fighters were valued at about 30 million dollars each, a big problem. Sure enough, in time, lawsuits were filed. Lockheed and the insurance companies agreed that the verdict guilty or not-guilty on the deaths would apply to the loss of the F-15 planes as well, thus saving the cost of two separate trials. This made it all the more critical to defend the death claims. When Tom Roe called me about this crash I advised him to, "keep his hands off or they would get burned." I knew the facts of the case were that an illegal part had been used to hold the control column in a natural position while the airplane was parked without hydraulic pressure. Lockheed had issued a safety bulletin telling operators not to restrain the control column with no hydraulic pressure. The crew forgot to remove the restraint before take-off even though the copilot responded to the check list challenge that the control column was "free and clear." I explained the facts as I knew them to Tom Roe and he actually followed my advice. However, he did play around the edges as a paid consultant to the plaintiff's attorneys "as he was experienced with C-130 suits."

After a lot of maneuvering, deposing, etc., the trial was finally set in Miami in the Federal Court. I knew that it would be long and drawn out so Louise and I took the motor home and set up in Homestead, Florida. From there I could take the Miami

Rapid Transit system and get off across the street from the local attorneys' offices after about an hour ride on the transit system. I was provided an office with a desk phone, computer connection, etc. in their building, which was half a block from the Federal Courthouse and settled in for the long haul. At one point, the Lockheed attorney handed me a note that said, "Get lost, the plaintiffs are trying to serve you with a subpoena and then treat you as a hostile witness." I took off and was glad that no one knew where we were staying.

Two days later the word arrived that they had rested their case and to come on back. There were a few snide remarks but they were finished. The process server caught up with me the following day and served me. I simply handed the subpoena to the Lockheed attorney who looked it and threw it in a waste basket. I spent two and a half days on the stand between direct and cross-examination. Following closing arguments, the jury went out and after about four hours deliberation returned with a verdict that Lockheed was not guilty on all counts.

I thought that was that, but up jumped the devil. The flight crew had been driven from the crew lounge to the airplane by a civilian female employee of the Air Force in an Air Force step van. She sued Lockheed because the "traumatic effect on her for driving that crew to their deaths had turned her into a nymphomaniac." I couldn't believe it, but the insurance company settled with her for $40,000, which was a lot cheaper than a trial. I was never contacted to do any work on her case, which did put it all to rest.

I had gone to the Lockheed Legal Department and the insurance company with the argument that each airplane was built to military specifications and standards, inspected by USAF inspectors over many days, flown by USAF test pilots, built under USAF Quality Control Inspectors' observation, and eventually certified as in compliance with the approved model specifications. Therefore, the notion of "design deficiency" was a lot of BS. I was told to perfect that argument and they would see how it flew. After about a year and a half of thinking, talking, and writing I went to Lockheed's attorneys and consulted with them.

They put it in legalize and then got it before the U.S. court. Believe it or not, eventually, the U.S. Supreme Court ruled in our favor. That ended suits against the airplane for "design deficiencies." Every once in a while some "hot shot" lawyer circumvents this ruling on a technicality like "it not being in the flight handbook." But to date none of these such suits have amounted to more than being a nuisance. What is amazing is that Boeing jumped on this ruling and got it expanded to the point that "design deficiency" for military designed planes is no longer grounds for a lawsuit.

The last time I was contacted regarding a C-130 accident was in September of 1998 by a Lockheed attorney defending a crash that occurred on November 22, 1996. I had been aware of the crash details and had done some investigating for Lockheed's insurance company. As a result I knew most of the facts of the accident and had a meeting with the attorney defending Lockheed. This suit was really not against the airplane but Lockheed's involvement with the instructions in the flight handbook. A couple of months later I received a fax from him saying that the case had been settled on "very favorable terms for Lockheed." Thus ended 44 years of my involvement with the Herc, which is still manufactured and flown today.

Epilogue

You name it and the Hercules "has been there and done that." The C-130 has been flown from both poles, landed or air-dropped cargo at every hot spot from the Congo to Vietnam to Kosovo to Afghanistan and Iraq, and hauled relief supplies to every outpost in the world. It has been used to airdrop 15,000 pound bombs, paratroopers, and leaflets that weigh only a few grams. The C-130 serves as a gunship, monitors and jams enemy radio transmissions; it is used to track icebergs in the North Atlantic and drug traffickers in the Caribbean and Pacific.

The Hercules flies into hurricanes to obtain wind and rain data; it is used to drop retardant on forest fires and insecticide on mosquito infestations. The C-130 has been used to transport cows, whales, elephants, horses, camels and everything in between. It has been used to medevac thousands of casualties to hospitals. A four-plane aerial demonstration team named The Four Horseman flew C-130s in aerial maneuvers demonstrating their maneuverability. A C-130 once carried 452 people in an emergency despite being designed to carry only 90.

There have been five major military versions of the C-130 along with around 70 special purpose variants. Between 1954 and 1959, 231 C-130As were built. Production of the C-130B ran from 1958 until 1963 and resulted in 230 aircraft. A total of 491 C-130Es were built from 1961 to 1974. The most produced version of the Hercules so far is the C-130H, with 1,205 aircraft coming off the assembly between 1964 and 1997. Production of the L-100, the civilian variant, totaled 115 aircraft and production ran primarily from 1964 to 1987. Talk about a versatile airplane design.

My thoughts go back to what I thought when I was first shown a three view drawing of the Herc as I was being assigned to Dept 72-05. I thought, "you can't sell that thing," and I was in good company. Kelly Johnson was the legendary head of the Lockheed "Skunk Works" at Lockheed California. He secretly headed up the design of the Lockheed P-80 Shooting Star which was the F-80 first USAF jet fighter, the F-104 interceptor which was the first production plane that could fly faster than the speed of sound, the U-2 high altitude spy plane and the SR-71 spy plane that holds the world's record for speed and altitude for a production plane.

When Kelly looked at the prototypes of the C-130 he said, "well, you built two and flooded the market."

C-130 Hercules airplanes fly all over the world today with approximately 2,300 having been built and delivered to more than 70 different operators in over 60 nations. That is the longest production run of any airplane in aviation history and it continues to this day at the original plant in Marietta, Georgia.

Am I proud of my involvement with the Herc? You can bet on it!

Ross Holdeman
U.S. Army Air Corps Flying Cadet 1942
Author 2008

APPENDIX

Enclosure (2)

PACIFIC WESTERN AIRLINES

YOUR FILE 10.01.47   OUR FILE

Terminal Building,
Industrial Airport,
Kingsway Avenue,
Edmonton, Alberta.

22 August, 1968.

Mr. D.T. Crockett,
Vice President,
Lockheed-Georgia Company,
Marietta, Georgia 30060.

Dear Sir:

  I was very pleased that your Mr. Ross Holdeman paid us a visit on August 2nd in regards to the engine problems we encountered a few days previously. This action permitted us to make a prompt decision and certainly cleared some points that were in doubt.

  Likewise, it was fortunate that Mr. Holdeman was here when we discovered the cracked upper wing plank. Again, his co-operation and action permitted a repair scheme to be developed and the necessary repair parts were quickly on their way.

  I wish to express my thanks to Mr. Holdeman and to the Lockheed-Georgia Company for their efficient and excellent service in these instances.

Yours very truly,

PACIFIC WESTERN AIRLINES LTD.

G. L. Cannam,
Maintenance Manager, N.O.

GLC/vlr

RECEIVED SEP 16 1968 J. W. CURRIE, JR.

D.T. CROCKETT
AUG 26 1968

ROSS HOLDEMAN

13 September 1968

Mr. G. Cannam
Maintenance Manager, N.O.
Pacific Western Airlines
Terminal Building - Industrial Airport
Kingsway Avenue
Edmonton, Alberta, Canada

Dear Mr. Cannam:

Thanks very much for your letter of 22 August and for the nice words about Ross Holdeman. It's always good to hear our customers speak well of our people and our service. We value highly PWA as a customer and we were pleased to be of service on the recent problems.

We feel very fortunate in having Ross on the Lockheed-Georgia team. As the Hercules Program Manager, I feel that customer service and satisfaction is key to future sales successes. The keen understanding of customer needs, plus personal initiative that Ross has, contributes greatly toward this end.

May I take this opportunity to say that we wish PWA continued success with the Hercules and that someday in the future we will have the pleasure adding another airplane to your operation.

Sincerely,

D. T. Crockett, Jr.

DTC:ct

cc: Ross Holdeman

My Life with the C-130

LOCKHEED-GEORGIA COMPANY
A DIVISION OF LOCKHEED AIRCRAFT CORPORATION
INTERDEPARTMENTAL COMMUNICATION

TO: R. M. Holdeman	DEPT. 72-05	ZONE 56	DATE October 17, 1968
FROM: A. E. Goode	DEPT. 72-05	ZONE 56	EXT. 2956

SUBJECT: COMMENDATION

The purpose of this IDC is to commend you for the outstanding manner in which you have represented the Engineering Department in the performance of the many and varied tasks associated with customer support activities on the L-100 airplane.

Without exception, each task to which you have been assigned has been pursued tenaciously and aggressively to a final and satisfactory conclusion, no matter how lengthy or complex the problem. Toward this end you have devoted not only your characteristic brand of vigor, but much of your own time as well. In particular, I want it to be noted for the record that essentially all of the direct customer support activities in which you have been involved have been carried out in the middle of the night, on weekends, and at other times and hours. Still, you have always responded to these customer calls for help, not with reluctance, but with enthusiasm. Also, you have somehow managed not only to minister to the customer's need at night, but have likewise managed to show up promptly the next morning to attend to your normal daytime duties.

It should also be noted that at your own insistence all of the customer support activities have been initiated on the basis of instantaneous, drop-everything, kind of response which has usually resulted in the need for you to travel to and from field assignments on your own time. This has not only resulted in reduced travel expense for the Company, but has also contributed significantly to our favorable image in the eyes of the customer, who psychologically equates a bright-and-early arrival time with Lockheed's desire to render superior service.

Although I could cite many specific examples of the kind of performance noted above, two will suffice to illustrate the spirit which characterizes your work. These are outlined in Enclosure (1).

Two other types of activity also deserve comment - 1) your work with the FAA, and 2) your handling of power plant and propeller problems.

Commendation   Page 2   October 17, 1968
R. M. Holdeman

In connection with FAA activities, your comprehensive knowledge of the Federal Aviation Regulations, plus a thorough understanding of how the entire regulatory system works, has been of invaluable aid in Lockheed's negotiations with both the Engineering Branch and the Air Carrier Section. Your success in these negotiations is proof of the respect in which your judgment and integrity is held by those with whom you have to deal - and "success", in at least one instance I can recall, has meant the lifting of a grounding order which otherwise would have been given.

As regards the power plant and propeller area, it is recognized by our customers, by the FAA, by the cognizant vendor representatives, and by knowledgeable Lockheed personnel, that you have no peer when it comes to detailed and intimate familiarity with these two systems. Your advice in this area of specialization is widely sought, not only in support of the L-100 program, but also in connection with many other programs where down-to-earth evaluations are needed. The fact that you are a rated and current C-130/L-100 pilot significantly enhances the credibility of your opinions on operational problems, and your past liaison experience and wrench-in-hand expertise on the maintenance stand lend credibility to your hardware decisions.

In light of all of the above I want to say "thanks" for a job well done, and again to commend you for the kind of performance which all of us should strive to achieve in order to bring to fruition the hope we share in the success of the L-100, an airplane of which you have frequently said you are proud, and an airplane in which we have much faith.

A. E. Goode
Assistant Project Engineer
C-130/C-141 Project Engrg. Division

AEG:sc   Approved: J. W. Flournoy

cc: D. T. Crockett   Project Engineer
    J. W. Currie   C-130/C-141 Project Engrg. Division
    F. N. Dickerman
    Personnel Folder

Enclosures: (1) ASA/PWA Service Coverage
            (2) PWA Letter, File 10.01.47

My Life with the C-130

Enclosure (1): ASA/PWA Service Coverage

MLG GEARBOX PROBLEM AT ALASKA AIRLINES

On this occasion you had the foresight to take spare parts with you. This required the negotiation for their release and clearance from the plant on a short-notice basis. With minimum delay, you reported to the ASA hangar in Seattle, set up and conducted a sophisticated system functional test, and successfully verified the problem. With the new spares you corrected the problem and then negotiated release of the airplane with the FAA. In this instance, you worked around the clock for three straight days. Then, upon your return to the plant, you arranged for further bench tests of the defective ASA parts, supervised a teardown inspection, and then made a trip to the vendor's facility to negotiate corrective action. You even authored the Engineering Report (ER-9933) which was used to finalize FAA approval of the various actions which had been taken.

PWA ENGINE AND WING CRACK PROBLEMS

This incident developed while you were at the Allison plant in Indianapolis to witness a teardown inspection of a failed DAL engine. As soon as you learned of PWA's engine problem you proceeded at once to PWA's Vancouver office, and then went on to meet the airplane at Edmonton, Alberta. After fixing the engine problem, you were diverted from your return trip by a report of a wing crack on the same airplane. You remained in Edmonton, surveyed the damage, and reported details of the problem to the Project Stress Group. From this a quick fix was devised, and parts were immediately fabricated and dispatched to PWA. In the meantime, you had supervised preparation of the airplane to receive the fix. Again, you worked around the clock for several days, but despite exhaustion, as soon as the airplane could be entrusted to other hands, you immediately began the return trip home in order to satisfy a Navy commitment for the following day in Washington. For this effort, on behalf on PWA, the Enclosure (2) letter of thanks was received by Mr. D. T. Crockett. It is also widely believed that as a result of the Washington trip, mentioned above, you were also able to salvage the T-4 program from being abandoned by the Navy for lack of a suitable approach to the problem of upgraded engine performance. This program, which is expected to be approved within the next few days, can potentially result in over one hundred million dollars worth of additional Navy business for Lockheed.

ROSS HOLDEMAN

LOCKHEED-GEORGIA COMPANY
A DIVISION OF LOCKHEED AIRCRAFT CORPORATION
INTERDEPARTMENTAL COMMUNICATION

TO: R. M. Holdeman   DEPT. 72-05 ZONE 458 DATE January 10, 1972

FROM: L. O. Kitchen   DEPT. 93-01 ZONE 46 EXT. 4-2941

SUBJECT: LETTER OF APPRECIATION

Ross, I have been given a report of two recent Delta incidents involving our L-100-20 aircraft. Your handling of the incident of December 27, in which you were working with Clayton Windsor, certainly deserves a word of appreciation and that is the purpose of this letter.

Delta is a valued customer and one to whom our best efforts must continue to be directed. You obviously recognized this, did a great job, and enhanced the Lockheed image of concern for the customer. My sincere thanks for your interest and performance.

L. O. Kitchen

LOK:rc

My Life with the C-130

LOCKHEED - GEORGIA COMPANY
A DIVISION OF LOCKHEED AIRCRAFT CORPORATION
MARIETTA   LOCKHEED   GEORGIA

D.T. CROCKETT, JR.
VICE PRESIDENT

March 28, 1972

Mr. R. M. Holdeman, Jr.
512 Allen Road, N. E.
Atlanta, Georgia 30324

Dear Ross,

Thanks for the fine job you did for the Italians in getting their C-130 fixed in Bermuda on 25/26 March 1972.

Ross, I know you left Atlanta on a 3:00 a.m. flight Saturday and were in the Bahamas prior to noon. I also know you started to "trouble shoot" the engine right away, found the problem and immediately called Earl Rainwater and told him to cancel the JetStar and back-up personnel. I know you personally worked on the engine until 11:00 p.m. Saturday and the aircraft was able to depart for Pisa Sunday morning.

Your keen understanding of the importance of prompt customer support to Lockheed's image was confirmed by your actions. I sincerely hope you will express Gelac's apologies to your family for causing you to be away over the weekend, and thank them for their understanding.

Your technical skill, loyalty and dedication to Lockheed is truly appreciated.

Sincerely,

D. T. Crockett, Jr.

DTC:ct

cc: L. O. Kitchen
    R. B. Ormsby
    J. J. Cornish, III

Ross Holdeman

Air HQ/32006/1/Acft.Engg.

Air Headquarters,
Peshawar

23rd October, 1972

Dear Sir,

VISIT OF LOCKHEED TECHNICAL REPRESENTATIVE

This is to thank you for arranging the visit of Mr. R. Holdeman of Lockheed Georgia Company to P.A.F. Chaklala to sort out maintenance problems being experienced on our C-130 aircraft.

This visit has proved very useful from our point of view. It provided us an opportunity to discuss in great detail our problems. Mr. Holdeman was able to suggest some remedial measures on the spot and also promised to send more information on his return to the Company. He also agreed to keep us informed about the latest developments.

We would be grateful if you could convey our appreciation to M/s Lockheed Georgia Company, U.S.A. for sending Mr. Holdeman to Pakistan at our request.

Thanking you,

Yours faithfully,

(M. TAQI)
Group Captain,
for Chief of the Air Staff
Pakistan Air Force

Mr. Aziz Jamal,
  A.R.O. Limited,
    43-N, Block 6,
      P.E.C.H.S.,
        Karachi - 29

A.R.O. Limited

43-N, Block 6, P.E.C.H.S., Karachi 29, Pakistan. Phone: 412444.
Cable: "Aviquipo."

AJ-8793.
October 28, 1972.

Mr. Ross M. Holdeman, Dept 72-05/458,
Engineering Specialist - Hercules,
Lockheed-Georgia Company,
86, South Cobb Drive,
MARIETTA. GEORGIA. 30060. (USA)

Dear Ross;

It gives me pleasure to forward to you a letter just received from Air Headquarters. Let me also add my apprecition of the time and efforts spent by you during your brief visit. I do trust that at some future date, you will have a little time to relax and see a bit of our country.

The family joins me in sending you our warmest regards.

Sincerely yours,

AZIZ JAMALL

AJ:SAAS.
Encl: The letter.

A.R.O. Limited

43-N, Block 6, P.E.C.H.S., Karachi 29, Pakistan. Phone: 41244.
Cable: "Aviquipo."

AJ-8794.

October 28, 1972.

Group Captain M. Taqi, T.Pk.,
Director of Aircraft Engineering,
Pakistan Air Force,
Air Headquarters,
PESHAWAR.

Sub: **VISIT OF LOCKHEED TECHNICAL REPRESENTATIVE**

Dear Group Captain;

A short note to acknowledge your letter No. Air HQ/32006/1/Acft.Engg dated 23rd October 1972, which I have conveyed to Lockheed and I am sure Mr. Holdeman will be most appreciative. We have been assured that he is actively engage in compiling the information which he promised to send us.

This will be forwarded to you as soon as received.

Once again with personal regards, I remain,

Yours sincerely,

AJ:SAAS.

AZIZ JAMALL

b.c.c. Mr. Ross M. Holdeman,

My Life with the C-130

DEPARTMENT OF THE AIR FORCE
TUSLOG DETACHMENT 175 (MAC)
APO NEW YORK 09289

REPLY TO
ATTN OF: Det 175/CC

SUBJECT: Letter of Appreciation

TO: R.M. Holdeman

1. I would like to say thanks for your hard work and support during the accident investigation. As you well know it was a difficult job accomplished under less than ideal conditions. In fact there was nothing concerning this accident that was easy or could be conducted by the book. It could be said that the mountains were tall, the walks were long, the mud and snow were deep, and the wreckage was heavy. You helped put a symbalance of order to the different aspects of the investigation.

2. It takes someone involved in the accident investigation to fully appreciate your efforts. You did much more than just analyze the wreckage and provide your expert opinion. Without your extra effort and much needed assistance we could still be on the mountain side.

3. Again, I say thanks. I felt quite comfortable briefing this accident at MAC Headquarters because of the assistance and analysis you provided the board. As you have probably heard, there is some disagreement with the initial cause but there is no evidence to support the disagreement. It was a pleasure working with you and I hope to see you again under more pleasant circumstances.

Wesley W Bean Jr.

WESLEY W. BEAN, JR., Colonel, USAF
Commancer

(TURKEY)

GLOBAL IN MISSION — PROFESSIONAL IN ACTION

Ross Holdeman

GEN/BRIG RAUL R. MORALES

PARTICULAR

26 October 1976

R.D. ORMSBY
PRESIDENT - LOCKHEED GEORGIA CO.
SOUTH COBB DR.
MARIETTA, GA 30063

Gentleman:

    This Headquarters wishes to acknowledge the commendable work of Mr. Ross Holdeman and Mr. Ted Faber in assisting us with the investigation of our C-130 loss of 03 September 1976. Their positive attitudes and professional assistance were most appreciated by our investigative personnel.

    Please convey our formal thanks to these fine representatives of your company.

RAUL RAMON MORALES, Brigadier General, FAV
Inspector/General.

LOCKHEED - GEORGIA COMPANY
A DIVISION OF LOCKHEED AIRCRAFT CORPORATION
MARIETTA, GEORGIA 30063

Robert B. Ormsby
PRESIDENT

November 9, 1976

Mr. R. M. Holdeman
512 Allen Road, N.E.
Atlanta, Georgia 30324

Dear Ross:

I am happy to enclose a copy of a letter from Brigadier General Raul Ramon Morales commending your work with the Venezuelan Air Force in the investigation of the loss of their C-130 on the 3rd of September. It is always gratifying to hear good reports on our employees' performance while representing the Company in other areas, and I would like to add my thanks to you for your obviously professional conduct in Venezuela.

I hope your current tour to the many countries operating the C-130 has been enjoyable and that it will be beneficial to you in the future.

Sincerely,

R. B. Ormsby

cc: W. P. Frech
R. D. Roche

UNITED STATES MARINE CORPS
Marine Aerial Refueler Transport Squadron 252
Marine Aircraft Group 14
2d Marine Aircraft Wing, FMFLant
Marine Corps Air Station, Cherry Point, N. C. 28533

23 January 1978

Mr. R. B. Ormsley
President
Lockheed-Georgia Company
86 South Cobb Drive
Marietta, Georgia 30063

Re: Mr. Leo Sullivan
    Mr. Ted Faber
    Mr. Ross Holdeman
    Mr. Dick Wilson

Dear Mr. Ormsley,

    I wish to express my personal gratitude as well as that of the officers and men of VMGR-252 to the Lockheed Georgia Company for allowing Mr. Leo Sullivan and team, Mr. Ted Faber, Mr. Ross Holdeman, and Mr. Dick Wilson, to present the C-130 Safety Briefing which we were privileged to enjoy on 18 and 19 January 1978. The entire program was technically enlightening and well presented. Arrangements for the team visit were handled by Mr. Thomas Morgan, our Lockheed Georgia Service Representative.

    It is not frequent enough that we users in the field have the opportunity to freely interchange ideas with such a team of experts. This form of communication is in part responsible for the over two hundred and ten thousand accident free flight hours VMGR-252 has accumulated over the past eighteen and one half years in the Lockheed KC-130 Hercules.

    Again, I thank you on the behalf of this command for sharing Messers, Sullivan, Faber, Holdeman, and Wilson with us.

B. R. BRIDGEWATER
Lieutenant Colonel, U. S. Marine Corps
Commanding

My Life with the C-130

LOCKHEED - GEORGIA COMPANY
A DIVISION OF LOCKHEED AIRCRAFT CORPORATION
MARIETTA, GEORGIA 30063

ROBERT B. ORMSBY
PRESIDENT

February 8, 1978

Mr. Ross M. Holdeman
512 Allen Road, N. E.
Atlanta, Ga. 30324

Dear Ross,

I have received a highly complimentary letter from Lt. Col. B. R. Bridgewater, Commander VMGR-252, about the briefing you and other Lockheed-Georgians gave at Cherry Point, January 18 and 19.

My thanks and commendation on a fine job which has benefited both Lockheed and a customer. The reception given this visit speaks strongly in support of continuing such meetings between us from the plant and those who fly and maintain our aircraft on a regular basis.

I appreciate the efforts you have made and am pleased by your commendation from Lt. Col. Bridgewater.

Sincerely,

Robert B. Ormsby

ROSS HOLDEMAN

LOCKHEED - GEORGIA COMPANY
A DIVISION OF LOCKHEED CORPORATION
MARIETTA, GEORGIA 30063

ROBERT B. ORMSBY
PRESIDENT

March 31, 1980

Mr. R. M. Holdeman, Jr.
Lockheed-Georgia Company
Department 72-05, Zone 80
Marietta, Georgia 30063

Dear Ross,

    Your fine performance has again brought credit to Lockheed-Georgia and to yourself.

    I have received a letter of appreciation from Col. T. W. A. Stuart, USAF, Commander of the Headquarters 1st Special Operations Wing (TAC) at Hurlburt Field, Florida. He writes to commend the team of Lockheedians -- Leo Sullivan, Ted Faber, Dave Valley, and Ross Holdeman -- who briefed the flight crews and maintenance people of his command on last February 14 and 15.

    "The wealth of knowledge gleaned from the presentations will enhance the safety program and was well received by all attendees," Colonel Stuart wrote.

    Thank you for another professional job well done. Your efforts in these personal contacts mean much to all of us and help keep Lockheed-Georgia out front. I appreciate the extra efforts you make.

                                  Sincerely,

                                  R. B. Ormsby

RBO:ed

MY LIFE WITH THE C-130

DEFENSE LOGISTICS AGENCY
DEFENSE CONTRACT ADMINISTRATION SERVICES REGION, LOS ANGELES
11099 SOUTH LA CIENEGA BOULEVARD
LOS ANGELES, CALIFORNIA 90045

IN REPLY
REFER TO   DCRL-QD

15 June 84

Mr. Ross M. Holdeman
Dept. 72-05 Zone 80
Lockheed Georgia Company
Marietta, GA 30060

Dear Mr. Ross ~~Holdeman~~:

On the occasion of your retirement, in the presence of your many friends and co-workers, I would like to add my congratulations and best wishes. I regret that last minute changes prevent my attendance to such a gathering.

Many great stories have probably been told, and I could make my contribution as well, but I hope that none present loose sight of their true significance. Such a wealth of remembrances can only occur when one has done much, contributed to many, and has as true result, earned the respect and admiration of those present.

You have, over the many years of our friendship and working together, enriched my life beyond my ability to express. You have been a patient teacher, loyal advisor, guardian of "the mission", caretaker of the flock, and a great person to share time with. We all owe you a great deal, which we probably will never be able to repay.

You leave a legacy of contributions to aviation, safety, and especially the many who have operated the C-130 throughout the world. Through you, a large portion of the much enjoyed Lockheed reputation has been planted, nourished, and harvested. You have been a good gardener!

Around you today you have many friends who represent but a small number of the many who have much to thank you for. We will miss you greatly.

Good luck and God speed!

Sincerely,

JERRY G. VALENTINE
Lieutenant Colonel, USAF
Deputy Director, Quality Assurance

Look to DCASR, Los Angeles for Leadership

Milton Keynes UK
Ingram Content Group UK Ltd.
UKHW021544210924
448546UK00004BA/30